Analytical Techniques in Electromagnetics

Analytical Techniques in Electromagnetics

Matthew N. O. Sadiku • Sudarshan R. Nelatury

CRC Press
Taylor & Francis Group
Boca Raton London New York

CRC Press is an imprint of the
Taylor & Francis Group, an **informa** business

CRC Press
Taylor & Francis Group
6000 Broken Sound Parkway NW, Suite 300
Boca Raton, FL 33487-2742

First issued in hardback 2019

ISBN-13: 978-1-4987-0901-9 (hbk)

Visit the Taylor & Francis Web site at
http://www.taylorandfrancis.com

and the CRC Press Web site at
http://www.crcpress.com

Contents

List of Figures

List of Tables

Preface

With the advent of computers and the accompanied explosion of commercial software packages, numerical methods in electromagnetics (EMs) began gaining the academic limelight. While numerical methods permit arbitrary shaped geometries in the problem domain, the user is still held responsible for accurate formulation of EM problems. Also, in order to validate the accuracy of numerical methods, one needs the knowledge of exact solutions of some canonical geometries as test cases. Numerical methods can be crisply placed under various categories such as method of moments (MoM), finite difference time domain (FDTD) method, finite element method (FEM), method of lines (MoL), and so on.

Over the years, a number of analytical and numerical techniques have been developed to deal with EM problems. Analytical methods provide exact, rigorous solutions to mathematical problems compared with numerical methods that provide approximate solutions. Although in many practical cases an analytical solution cannot be obtained and we must resort to numerical approximate solution, an analytical solution is useful in verifying solutions obtained from numerical methods. Therefore, analytical and numerical techniques should be considered as complementary.

Analytical solutions are fairly helpful and are always preferred for the following reasons: (1) They provide exact solutions which may be used to validate the accuracy of approximate solutions. (2) A better insight of the actual physical process is easily gained. (3) They can be utilized in finding the precise quantity of interest over a whole range of parameters' values.

This book is organized into six chapters. With a short review of basic EM in Chapter 1, we start with the discussion of separation of variables technique in Chapter 2. When it is possible to frame the problem geometry in one of the 11 orthogonal coordinate systems, this method is applicable. We shall describe the use of this method in case of the Laplace's equation, heat equation, and the wave equation. The three standard coordinate systems such as rectangular, cylindrical, and spherical are dealt with. In Chapter 3, the series expansion method is presented. The well known Fourier series method is taught even in undergraduate courses like signals and systems. The technique of series expansion in this chapter is the same undertone involving general orthogonal sets of functions used to obtain the analytical solutions in the connotation of EM boundary value problems. Specifically, we provide the solution of Poisson's equation in a cube and in a cylinder, and also show scattering by cylinders and spheres as simple examples.

Chapter 4 deals with the conformal transformation technique. When an EM transmission structure is uniform in the longitudinal direction, the important primary constant, which is capacitance per unit length, can be obtained by mapping the cross-sectional geometry to a known shape that is easy to solve. The fact that the real and imaginary parts of an analytic function of a complex variable satisfy Cauch–Riemann equations and individually are solutions of the Laplace equation underlies the notion of complex potential function. The real and imaginary parts of it can be used to represent the potential and flux functions which are orthogonal to each other. When we know the analytical expressions for the real and the imaginary parts, it is possible to sketch equipotential and flux lines. A visual display of these diagrams is very helpful to the reader to gain insight into the physical aspects. We have provided a good collection of examples and shown the field plots. Of particular interest is the Schwarz–Christoffel transformation useful in analyzing strip and microstrip lines. We gave a brief introduction to this and showed examples.

Just as the phasor method is convenient in doing sinusoidal steady state analysis in circuits replacing calculus with simple algebra, integral transform techniques are powerful tools that convert partial differential equations into simpler ordinary differential equations by transforming one spatial variable into its spectral domain. The type of transform and its kernel depend on the geometry. Several transforms are proposed in the literature. In Chapter 5, we include Fourier sine and cosine transforms, two-sided Fourier transform, Laplace transform, Hankel transform, and the Mellin transform. The use of these transform techniques is illustrated with few worked out examples.

Chapter 6 deals with perturbation techniques. Often we come across an EM structure whose shape is slightly different from one whose solution is well established, or there could be a change in the material media filling the structure. When there is thus a shape perturbation or material perturbation, one could obtain the solution in the form of a series involving the powers of the perturbation parameter. This approach offers results that are of reasonable accuracy. It is intuitively satisfying to see the perturbed results degenerate to their unperturbed version as the perturbation parameter tends to zero. We have furnished some interesting examples in this chapter.

This monograph is designed for researchers, scientists, and engineers seeking analytical solutions to EM problems. It is important to keep a balanced view of techniques for solving EM problems. To overemphasize the importance of analytical methods at the expense of numerical techniques would not reflect the trends in technology. Therefore, this book does not cover all possible analytical methods and does not cover the methods in a great depth like a mathematical monograph. The topics have been carefully selected to give the readers an appreciation of the kinds of problems that can be solved exactly. It is hoped that the book is useful to practicing engineers and scientists who are analytically oriented.

Several of the problems both worked out and given at the end of chapters as exercises have been gleaned from the references cited. We might have slightly modified the notation, but due credit belongs to the authors of these works.

We wish to place our sincere thanks to our spouses, Kikelomo and Chandu, for their patience and cooperation in completing the manuscripts. Thanks to Finney and Wesley for their occasional help.

M. N. O. Sadiku
S. R. Nelatury

Authors

Matthew N. O. Sadiku earned his PhD from Tennessee Technological University, Cookeville, TN. From 1984 to 1988, he was an assistant professor at Florida Atlantic University, where he did graduate work in computer science. From 1988 to 2000, he was at Temple University, Philadelphia, PA, where he became a full professor. From 2000 to 2002, he was with Lucent/Avaya, Holmdel, NJ as a system engineer and with Boeing Satellite Systems as a senior scientist. He is presently a professor at Prairie View A&M University Prairie View, TX.

He is the author of over 240 professional papers and over 60 books including *Elements of Electromagnetics* (Oxford University Press, 6th ed., 2015), *Fundamentals of Electric Circuits* (McGraw-Hill, 5th ed., 2013, with C. Alexander), *Numerical Techniques in Electromagnetics with Matlab®* (CRC Press, 3rd ed., 2009), and *Metropolitan Area Networks* (CRC Press, 1995). Some of his books have been translated into French, Korean, Chinese (and Chinese Long Form in Taiwan), Italian, Portuguese, and Spanish. He was the recipient of the 2000 McGraw-Hill/Jacob Millman Award for outstanding contributions in the field of electrical engineering. He was also the recipient of Regents Professor award for 2012–2013 by the Texas A&M University System.

His current research interests are in the areas of numerical modeling of electromagnetic systems and computer communication networks. He is a registered professional engineer and a fellow of the Institute of Electrical and Electronics Engineers (IEEE) for contributions to computational electromagnetics and engineering education. He was the IEEE Region 2 Student Activities Committee chairman. He was an associate editor for IEEE Transactions on Education. He is also a member of Association for Computing Machinery (ACM).

Sudarshan R. Nelatury was born in Andhra Pradesh, India. He earned his MS and PhD from the Osmania University, Hyderabad, India in 1985 and 1996 respectively. From 1983 to 1999, he taught at the Electronics & Communication Engineering Department, Osmania University. From 1999 to 2003, he was a visiting faculty member at Villanova University, Villanova, PA. In 2003, he joined the School of Engineering, Penn State Erie, Erie PA, where he is currently an associate professor. During the Spring of 2012, while on sabbatical, he was a visiting faculty member at the School of Engineering and

Applied Science, University of Pennsylvania, Philadelphia, PA. His interests are in electromagnetics and signal processing. He authored over 100 professional papers and co-authored the solution manual and book chapters with Dr. Sadiku. He was the recipient of the Outstanding Research Award, from the School of Engineering, and The Council of Fellows Research Award from Penn State Erie, The Behrend College in 2008. Dr. Nelatury is a senior member of the Institute of Electrical and Electronics Engineers.

1

Review of Electromagnetics

Science without religion is lame; religion without science is blind.

– Albert Einstein

CONTENTS

Electromagnetics (EM) is that branch of physics or electrical engineering in which the interactions between charges at rest and in motion are examined. It entails the analysis, synthesis, physical interpretation, and application of electric and magnetic fields. EM theory is indispensable to the design and operation of many practical devices. EM principles find applications in diverse areas such as microwaves, antennas, electric machines, bioelectromagnetics, plasmas, fiber optics, satellite communication, electromagnetic compatibility, radar meteorology, and remote sensing. A detailed treatment of the physical laws governing electromagnetic phenomena can be found in standard introductory EM texts such as Refs. [1–4]. Only a brief review will be presented here.

1.1 Maxwell's Equations

The foundational basis of electromagnetic theory was established in 1873 by James Clerk Maxwell, who for the first time unified observations of Michael Faraday, Karl Friedrich Gauss, and Andre-Marie Ampere. In this chapter, we present Maxwell's equations in time-domain and frequency-domain and the constitutive equations. SI units are assumed throughout. At macroscopic level, electric and magnetic phenomena are described by Maxwell's equations. Maxwell's work was based on large body of theoretical and empirical works of others such as Gauss, Ampere, and Faraday. In differential form, Maxwell's equations are given by

$$\nabla \times \mathbf{E} = -\frac{\partial \mathbf{B}}{\partial t} \qquad \text{Faraday's law} \tag{1.1}$$

$$\nabla \times \mathbf{H} = \mathbf{J} + \frac{\partial \mathbf{D}}{\partial t} \qquad \text{Ampere's law} \tag{1.2}$$

$$\nabla \cdot \mathbf{D} = \rho_v \qquad \text{Gauss' law for electric flux} \tag{1.3}$$

$$\nabla \cdot \mathbf{B} = 0 \qquad \text{Gauss' law for magnetic flux} \tag{1.4}$$

where **E**, **H**, **D**, **B**, **J**, and ρ_v are real functions of space and time. Table 1.1 has the meaning of these quantities and their units.

Equations 1.1 through 1.4 are referred to as Maxwell's equations in differential (point or local) form. They are first-order linear coupled differential equations relating the vector field quantities to each other at any given point. In their equivalent integral (or global) form these equations are:

$$\oint_L \mathbf{E} \cdot d\mathbf{l} = -\frac{\partial}{\partial t} \iint_S \mathbf{B} \cdot d\mathbf{S} \qquad \text{Faraday's law} \tag{1.5}$$

$$\oint_L \mathbf{H} \cdot d\mathbf{l} = \iint_S \mathbf{J} \cdot d\mathbf{S} + \frac{\partial}{\partial t} \iint_S \mathbf{D} \cdot d\mathbf{S} \qquad \text{Ampere's law} \tag{1.6}$$

TABLE 1.1
Various Quantities Appearing in Maxwell's Equations

Symbol	Meaning	SI Units
E	Electric field intensity	Volts per meter (V/m)
D	Electric flux density	Coulombs per square meter (C/m^2)
H	Magnetic field intensity	Amperes per meter (A/m)
B	Magnetic flux density	Webers per square meter (Wb/m^2) or Tesla (T)
J	Electric current density	Amperes per square meter (A/m^2)
ρ_v	Electric charge density	Coulombs per cubic meter (C/m^3)

$$\oiint_S \mathbf{D} \cdot d\mathbf{S} = \iiint_v \rho_v \, dv \qquad \text{Gauss's law for electric flux} \quad (1.7)$$

$$\oiint_S \mathbf{B} \cdot d\mathbf{S} = 0 \qquad \text{Gauss's law for magnetic flux} \quad (1.8)$$

where v is an arbitrary volume bounded by a closed surface S in Equation 1.7 and S is an arbitrary surface bounded by a closed path L in Equations 1.5 and 1.6. Notice that the terms on the right hand side of Equations 1.1 through 1.8 represent the sources of the fields viewed as vector point functions. The integral form of Maxwell's equations depicts the underlying physical laws, whereas the differential form is used more frequently in solving problems. Maxwell's equations contain an important relationship between charge and current. Taking the divergence of both sides of Equation 1.2 and applying the vector identity $\nabla \cdot (\nabla \times \mathbf{A}) \equiv 0$ yields

$$\nabla \cdot (\nabla \times \mathbf{H}) = \nabla \cdot \mathbf{J} + \nabla \cdot \frac{\partial \mathbf{D}}{\partial t} = 0 \qquad (1.9)$$

Introducing Equation 1.3 leads to

$$\nabla \cdot \mathbf{J} + \frac{\partial \rho_v}{\partial t} = 0 \qquad (1.10)$$

which is the continuity equation expressing the conservation of charge and current. Another equation associated with Maxwell's equation is the Lorentz force law, which relates the electromagnetic fields to measurable forces. If a charge q moves in an electric field $\mathbf{E}$ and magnetic field $\mathbf{B}$ at a velocity $\mathbf{v}$, the force on the charge is given by Lorentz force law:

$$\mathbf{F} = q(\mathbf{E} + \mathbf{v} \times \mathbf{B}) \qquad (1.11)$$

When the charge is not discrete, Lorentz law for the force per unit volume becomes

$$\mathbf{F} = \rho_v \mathbf{E} + \mathbf{J} \times \mathbf{B} \qquad (1.12)$$

1.2 Constitutive Relations

Maxwell's equations contain more variables than equations. To solve an EM problem using Maxwell's equations therefore requires some additional set of relations between the field intensities $\mathbf{E}$, $\mathbf{H}$, and flux densities $\mathbf{D}$, $\mathbf{B}$. Thus, we need the constitutive relations to supplement Maxwell's equation. For an isotropic medium, the constitutive relations are

$$\mathbf{D} = \epsilon\mathbf{E} = \epsilon_0\epsilon_r\mathbf{E} \tag{1.13}$$

$$\mathbf{B} = \mu\mathbf{H} = \mu_0\mu_r\mathbf{H} \tag{1.14}$$

$$\mathbf{J} = \sigma\mathbf{E} \tag{1.15}$$

where

$\epsilon_0 = \dfrac{10^{-9}}{36\pi}$ = permittivity of free space in Farads per meter (F/m)

ϵ_r = dielectric constant (or relative permittivity) of the medium

ϵ = permittivity of the medium in Farads per meter (F/m)

$\mu_0 = 4\pi \times 10^{-7}$ = permeability of free space in Henries per meter (H/m)

μ_r = relative permeability of the medium

μ = permeability of the medium in Henries per meter (H/m)

σ = conductivity of the medium in Siemens per meter (S/m)

Equations 1.13 through 1.15 are called *constitutive relations* for the medium in which the fields exist, while ϵ , μ , and σ are known as the *constitutive parameters*. The constitutive parameters are often used to characterize the properties of a material. For example, a dielectric material is said to be *linear* if $\mathbf{D} = \epsilon\mathbf{E}$ applies and if ϵ does not change with the applied $\mathbf{E}$ field and superposition holds between the two vector fields; it is *homogeneous* if ϵ does not change from point to point; and it is *isotropic* if ϵ does not change with direction. A material is said to be *dispersive* if its constitutive parameters are functions of frequency. For anisotropic media, the constitutive relations become

$$\mathbf{D} = \overline{\overline{\epsilon}} \cdot \mathbf{E} \tag{1.16}$$

$$\mathbf{B} = \overline{\overline{\mu}} \cdot \mathbf{H} \tag{1.17}$$

where $\overline{\overline{\epsilon}}$ and $\overline{\overline{\mu}}$ are permittivity and permeability tensors, respectively. In this case, the vector field $\mathbf{D}$ is no longer parallel to $\mathbf{E}$ and the vector field $\mathbf{B}$ is no longer parallel to $\mathbf{H}$. For example, in Cartesian coordinates system, Equation 1.16 becomes

$$\begin{bmatrix} D_x \\ D_y \\ D_z \end{bmatrix} = \begin{bmatrix} \epsilon_{xx} & \epsilon_{xy} & \epsilon_{xz} \\ \epsilon_{yx} & \epsilon_{yy} & \epsilon_{yz} \\ \epsilon_{zx} & \epsilon_{zy} & \epsilon_{zz} \end{bmatrix} \begin{bmatrix} E_x \\ E_y \\ E_z \end{bmatrix} \tag{1.18}$$

For crystals,

$$\overline{\overline{\epsilon}} = \begin{bmatrix} \epsilon_x & 0 & 0 \\ 0 & \epsilon_y & 0 \\ 0 & 0 & \epsilon_z \end{bmatrix} \tag{1.19}$$

indicating that crystals are described by symmetric permittivity tensor.

1.3 Boundary Conditions

Along the interface where the media involved exhibit discontinuity in electrical properties or there exist sources along the interface, the field vectors themselves are discontinuous and their behavior across the interface is governed by the boundary conditions. In fact, the solution of Maxwell's equations is not complete until the boundary conditions are imposed. The boundary conditions at the interface separating two different media 1 and 2, with parameters $(\sigma_1, \epsilon_1, \mu_1)$ and $(\sigma_2, \epsilon_2, \mu_2)$ as shown in Figure 1.1 are easily derived from the integral form of Maxwell's equations.

They are

$$E_{2t} - E_{1t} = 0 \quad \text{or} \quad (\mathbf{E}_2 - \mathbf{E}_1) \times \mathbf{a}_{n_{12}} = \mathbf{0} \tag{1.20}$$

$$H_{2t} - H_{1t} = K \quad \text{or} \quad (\mathbf{H}_2 - \mathbf{H}_1) \times \mathbf{a}_{n_{12}} = -\mathbf{K} \tag{1.21}$$

$$D_{2n} - D_{1n} = \rho_s \quad \text{or} \quad (\mathbf{D}_2 - \mathbf{D}_1) \cdot \mathbf{a}_{n_{12}} = \rho_s \tag{1.22}$$

$$B_{2n} - B_{1n} = 0 \quad \text{or} \quad (\mathbf{B}_2 - \mathbf{B}_1) \cdot \mathbf{a}_{n_{12}} = 0 \tag{1.23}$$

where $\mathbf{a}_{n_{12}}$ is a unit normal vector directed from medium 1 to medium 2; subscripts 1 and 2 denote fields in regions 1 and 2; and subscripts t and n denote tangent and normal components of the fields. Equations 1.20 and 1.23 state that the tangential components of **E** and the normal components of **B** are continuous across the boundary. Equation 1.21 states that the tangential component of **H** is discontinuous by the surface current density K on the boundary. Equation 1.22 states that the discontinuity in the normal component of **D** is the same as the surface charge density ρ_s on the boundary.

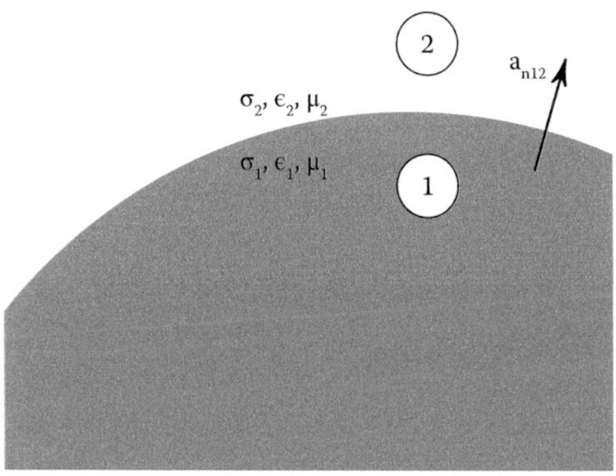

FIGURE 1.1
An interface between two media.

Equations 1.20 and 1.21 can be written in terms of the tangential components of the fields as

$$H_{2t} = H_{1t} + \hat{\boldsymbol{\tau}} \cdot (\mathbf{K} \times \mathbf{a}_{n_{12}}) \tag{1.24}$$

$$E_{2t} = E_{1t} \tag{1.25}$$

where $\hat{\boldsymbol{\tau}}$ is a unit vector tangential to the boundary along which the tangential field components are measured. If medium 1 is a perfect conductor, the boundary conditions in Equations 1.24 and 1.25 become

$$H_{2t} = \hat{\boldsymbol{\tau}} \cdot (\mathbf{K} \times \mathbf{a}_{n_{12}}) \tag{1.26}$$

$$E_{2t} = 0 \tag{1.27}$$

A boundary on which Equation 1.27 is valid is known as an electric wall since the tangential components of **E** are "shorted out" and must vanish at the surface of the conductor. Dual to this is the magnetic wall, which is a boundary on which **H** vanishes. For this case, assuming that it carries a *surface magnetic current* $\mathbf{M}_s$,

$$H_{2t} = 0 \tag{1.28}$$

$$E_{2t} = \hat{\boldsymbol{\tau}} \cdot (\mathbf{a}_{n_{12}} \times \mathbf{M}_s) \tag{1.29}$$

Finally, when dealing with problems that have one or more infinite boundaries, it is required that at an infinite distance from the source, the fields must be vanishingly small.

1.4 Power and Energy

Energy can be transported from one point to another by means of EM waves. (This is the case in wireless communication systems.) The rate of such energy transportation can be obtained from Maxwell's equations. Dotting both sides of Equation 1.2 with **E** and replacing **J** by $\sigma\mathbf{E}$ leads to

$$\mathbf{E} \cdot (\nabla \times \mathbf{H}) = \sigma E^2 + \mathbf{E} \cdot \epsilon \frac{\partial \mathbf{E}}{\partial t} \tag{1.30}$$

Using the vector identity

$$\nabla \cdot (\mathbf{A} \times \mathbf{B}) \equiv \mathbf{B} \cdot (\nabla \times \mathbf{A}) - \mathbf{A} \cdot (\nabla \times \mathbf{B}) \tag{1.31}$$

gives

$$\mathbf{H} \cdot (\nabla \times \mathbf{E}) - \nabla \cdot (\mathbf{E} \times \mathbf{H}) = \sigma E^2 + \frac{1}{2} \epsilon \frac{\partial E^2}{\partial t} \tag{1.32}$$

From Equations 1.1 and 1.14,

$$\mathbf{H}\cdot(\nabla\times\mathbf{E}) = \mathbf{H}\cdot\left(-\mu\frac{\partial\mathbf{H}}{\partial t}\right) = -\frac{1}{2}\,\mu\,\frac{\partial H^2}{\partial t} \tag{1.33}$$

Substituting Equations 1.32 into 1.33,

$$-\nabla\cdot(\mathbf{E}\times\mathbf{H}) = \sigma E^2 + \frac{1}{2}\,\epsilon\,\frac{\partial E^2}{\partial t} + \frac{1}{2}\,\mu\,\frac{\partial H^2}{\partial t} \tag{1.34}$$

Taking the volume integral of both sides and applying the divergence theorem to the left-hand side yields

$$\underbrace{-\oint\!\!\oint_S \mathbf{E}\times\mathbf{H}\cdot d\mathbf{S}}_{\text{Total power entering the volume}} = \underbrace{\frac{\partial}{\partial t}\iiint_v \left(\frac{1}{2}\,\epsilon\,E^2 + \frac{1}{2}\,\mu\,H^2\right) dv}_{\text{Rate of increase in the total energy in the EM fields in the volume}} + \underbrace{\iiint_v \sigma E^2 v}_{\text{Ohmic power loss in the volume}} \tag{1.35}$$

where volume v is bounded by the closed surface S. This expresses conservation of power or Poynting's theorem. The first term on the right-hand side is interpreted as the rate of increase in energy stored in the electric and magnetic fields. The second term is the power dissipated due to the conductivity of the medium ($\sigma \neq 0$). The quantity $\mathbf{E}\times\mathbf{H}$ on the left-hand side of Equation 1.35 is known as the Poynting vector $\mathbf{P}$ given by

$$\mathbf{P} = \mathbf{E}\times\mathbf{H} \tag{1.36}$$

$\mathbf{P}$ is the flow of power flux per unit area. It represents the power flux density and its units are watts/m^2.

1.5 Vector and Scalar Potentials

To avoid the difficulty involved in solving directly the vector fields associated with Maxwell's equations, it is sometimes expedient to use the electric scalar potential V and magnetic vector potential $\mathbf{A}$. The source for the vector potential is the current density and that of the scalar potential is the charge density. Note that the vector potential is generated by vector source and the scalar potential is produced by scalar source. From the solenoidal nature of the magnetic flux density, we choose to define the vector potential $\mathbf{A}$ in terms of the magnetic flux density:

$$\mathbf{B} = \nabla\times\mathbf{A} \tag{1.37}$$

But $\nabla \times \mathbf{E} = -\frac{\partial \mathbf{B}}{\partial t}$ so that

$$\nabla \times \left(\mathbf{E} + \frac{\partial \mathbf{A}}{\partial t}\right) = \mathbf{0} \tag{1.38}$$

Since the curl of the gradient of any scalar field is zero, the bracketed term in Equation 1.38 can be replaced by the gradient of a scalar potential V.

$$\left(\mathbf{E} + \frac{\partial \mathbf{A}}{\partial t}\right) = -\nabla V \tag{1.39}$$

or

$$\mathbf{E} = -\frac{\partial \mathbf{A}}{\partial t} - \nabla V \tag{1.40}$$

From Equations 1.37 and 1.40, we can determine the vector fields **B** and **E** provided the potentials **A** and V are known. By taking the divergence of Equation 1.40 and introducing Equation 1.3, we have

$$\begin{aligned} \nabla \cdot \mathbf{E} = \frac{\rho_v}{\epsilon} &= -\frac{\partial \nabla \cdot \mathbf{A}}{\partial t} - \nabla^2 V \\ \frac{\partial \nabla \cdot \mathbf{A}}{\partial t} + \nabla^2 V &= -\frac{\rho_v}{\epsilon} \end{aligned} \tag{1.41}$$

Taking the curl of Equation 1.37 and incorporating Equations 1.2 and 1.40 result in

$$\nabla \times \nabla \times \mathbf{A} = \mu \mathbf{J} - \mu\epsilon \nabla \frac{\partial V}{\partial t} - \mu\epsilon \frac{\partial \mathbf{A}^2}{\partial t^2} \tag{1.42}$$

Applying the vector identity

$$\nabla \times \nabla \times \mathbf{A} \equiv \nabla(\nabla \cdot \mathbf{A}) - \nabla^2 \mathbf{A} \tag{1.43}$$

leads to

$$\nabla^2 \mathbf{A} - \nabla\nabla \cdot \mathbf{A} = -\mu \mathbf{J} + \mu\epsilon \nabla \frac{\partial V}{\partial t} + \mu\epsilon \frac{\partial \mathbf{A}^2}{\partial t^2} \tag{1.44}$$

A vector field is uniquely defined when its curl and divergence are specified. The curl of **A** has been specified in Equation 1.37. We may choose the divergence of **A** as

$$\nabla \cdot \mathbf{A} = -\mu\epsilon \frac{\partial V}{\partial t} \tag{1.45}$$

This choice relates **A** and V and it is called the Lorentz gauge condition. By imposing this condition, Equations 1.41 and 1.44 become

$$\nabla^2 V - \mu\epsilon \frac{\partial V^2}{\partial t^2} = -\frac{\rho_v}{\epsilon} \tag{1.46}$$

and

$$\nabla^2 \mathbf{A} - \mu\epsilon \frac{\partial \mathbf{A}^2}{\partial t^2} = -\mu \mathbf{J} \tag{1.47}$$

which are wave equations. Once Equations 1.46 and 1.47 are solved, the fields **E** and **B** are obtained from Equations 1.37 and 1.40. The magnetic vector potential **A** does not have any physical significance just like the scalar potential V and is solely a mathematical convenience in determining EM fields.

Under static conditions, the electric and magnetic fields can exist separately. For electrostatic fields, the appropriate Maxwell's equations in Equations 1.1 and 1.3 become

$$\begin{aligned} \nabla \times \mathbf{E} &= \mathbf{0} \\ \nabla \cdot \mathbf{D} &= \rho_v \end{aligned} \tag{1.48}$$

In terms of the electric potential V (in volts), **E** can be expressed as

$$\mathbf{E} = -\nabla V \tag{1.49}$$

The relationship between the potential V and vector fields **E** and **D** is portrayed in Figure 1.2. Combining Equations 1.3 and 1.49 gives Poisson's equation:

$$\nabla \cdot \epsilon \nabla V = -\rho_v \tag{1.50}$$

or if ϵ is constant,

$$\nabla^2 V = -\frac{\rho_v}{\epsilon} \tag{1.51}$$

When $\rho_v = 0$, Equation 1.51 becomes the Laplace's equation:

$$\nabla \cdot \epsilon \nabla V = 0 \tag{1.52}$$

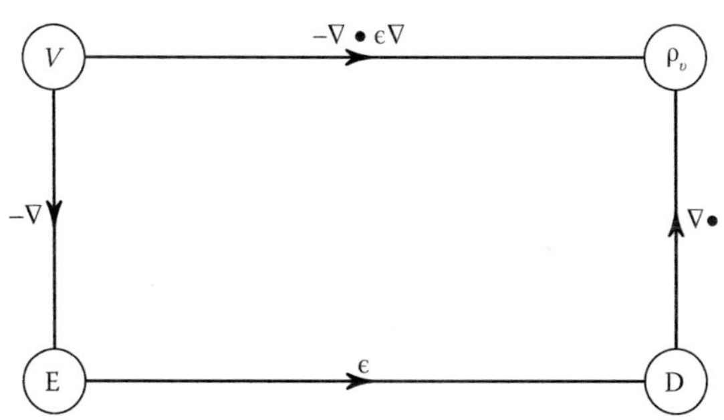

FIGURE 1.2
Relationship between V, **E**, and **D**.

or if ϵ is constant,

$$\nabla^2 V = 0 \tag{1.53}$$

Equations 1.50 and 1.52 are respectively Poisson's and Laplace's equations for inhomogeneous media, while Equations 1.51 and 1.53 are respectively Poisson's and Laplace's equations for homogeneous media. For magnetostatic fields, Maxwell's equations 1.2 and 1.4 become

$$\nabla \cdot \mathbf{B} = 0 \tag{1.54}$$

$$\nabla \times \mathbf{H} = \mathbf{J} \tag{1.55}$$

Using Equation 1.14 in Equation 1.55 we find that $\nabla \times \mathbf{B} = \mu\mathbf{J}$ which, from Equation 1.37, may be rewritten in terms of the magnetic vector potential $\mathbf{A}$ (in Weber/meter) as

$$\nabla \times \nabla \times \mathbf{A} = \mu\mathbf{J} \tag{1.56}$$

Applying the vector identity

$$\nabla \times \nabla \times \mathbf{F} \equiv \nabla(\nabla \cdot \mathbf{F}) - \nabla^2\mathbf{F} \tag{1.57}$$

to Equations 1.55 and 1.56 and assuming that $\nabla \cdot \mathbf{A} = 0$, we get Poisson's equation for magnetostatic fields:

$$\nabla^2\mathbf{A} = -\mu\mathbf{J} \tag{1.58}$$

When $\mathbf{J} = \mathbf{0}$, Equation 1.58 becomes Laplace's equation

$$\nabla^2\mathbf{A} = \mathbf{0} \tag{1.59}$$

Figure 1.3 presents the relationship between the vector potential $\mathbf{A}$ and vector fields $\mathbf{H}$ and $\mathbf{B}$.

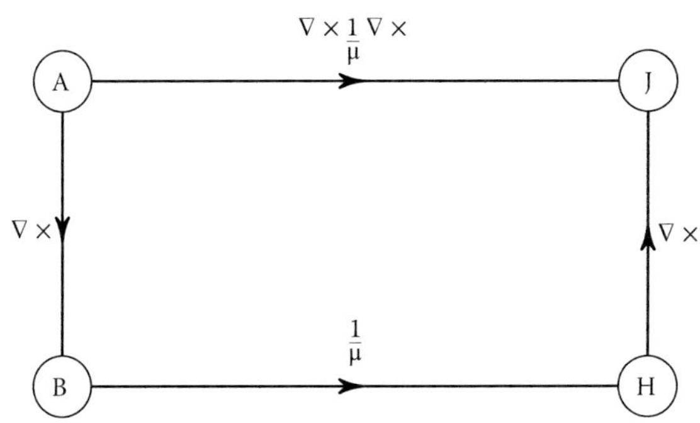

FIGURE 1.3
Relationship between **A**, **B**, and **H**.

1.6 Time Harmonic Fields

In many practical applications involving EM fields, the time variations are of sinusoidal form and are referred to as time-harmonic. Also, it is expedient to first understand the behavior of monochromatic waves (which have only a single frequency) and from them construct any other waveform using Fourier analysis. For time-harmonic fields, the time variation can be represented by $e^{j\omega t}$. Hence, for an arbitrary time-harmonic field $\mathbf{A}$ given by

$$\mathbf{A}(x, y, z, t) = \mathfrak{Re}\{\mathbf{A}_s(x, y, z, t)e^{j\omega t}\} \tag{1.60}$$

where $\mathbf{A}_s$ is the phasor form of $\mathbf{A}$ and is only a function of position in space (x, y, z). The symbol $\mathfrak{Re}$ stands for "Real part of". $\mathbf{A}$ in Equation 1.60 could stand for any of the EM fields $\mathbf{E}, \mathbf{H}, \mathbf{B}$, or $\mathbf{D}$. We notice from Equation 1.60, the differential and integral operations respectively yield

$$\frac{\partial \mathbf{A}}{\partial t} \rightarrow j\omega \mathbf{A}_s \tag{1.61}$$

$$\int \mathbf{A}dt \rightarrow \frac{\mathbf{A}_s}{j\omega} \tag{1.62}$$

Thus, for time-harmonic fields, Maxwell's equations can be written in simpler forms as

Differential form	Integral form	
$\nabla \times \mathbf{E}_s = -j\omega\mathbf{B}_s$	$\oint_L \mathbf{E}_s \cdot d\mathbf{l} = -j\omega \iint_{\mathbf{S}} \mathbf{B}_s \cdot d\mathbf{S}$	(1.63)
$\nabla \times \mathbf{H}_s = \mathbf{J}_s + j\omega\mathbf{D}_s$	$\oint_L \mathbf{H}_s \cdot d\mathbf{l} = \iint_{\mathbf{S}} \mathbf{J}_s \cdot d\mathbf{S} + j\omega \iint_{\mathbf{S}} \mathbf{D}_s \cdot d\mathbf{S}$	(1.64)
$\nabla \cdot \mathbf{D}_s = \rho_s$	$\oiint_{\mathbf{S}} \mathbf{D}_s \cdot d\mathbf{S} = \iiint_v \rho_{vs}\, dv$	(1.65)
$\nabla \cdot \mathbf{B}_s = 0$	$\oiint_{\mathbf{S}} \mathbf{B}_s \cdot d\mathbf{S} = 0$	(1.66)

Also, Maxwell's equations are easier to solve for time-harmonic fields. For time-harmonic fields, there are theorems that assure uniqueness of solutions if the boundary conditions on tangential fields are met. If the tangential components of the $\mathbf{E}$ and $\mathbf{H}$ fields satisfy the boundary conditions, the normal components of the same fields automatically satisfy their appropriate boundary conditions.

We can also show that for time-harmonic fields, the time-average Poynting vector is

$$\mathbf{P}_{ave} = \frac{1}{T}\int_0^T \mathbf{E} \times \mathbf{H}\, dt = \frac{1}{2}\mathfrak{Re}\,(\mathbf{E}_s \times \mathbf{H}_s^*) \tag{1.67}$$

where T is period of the sinusoidal variation and asterisk $*$ denotes complex conjugate.

1.7 Wave Equations

Consider a linear, isotropic, homogeneous, lossless dielectric medium ($\sigma = 0$) that is source free $\rho_v = 0, \mathbf{J} = \mathbf{0}$. We take the curl of both sides of Equation 1.1,

$$\nabla \times \nabla \times \mathbf{E} = -\mu \frac{\partial}{\partial t} \nabla \times \mathbf{H} \tag{1.68}$$

With $\mathbf{J} = \mathbf{0}$ Equation 1.1 reads,

$$\nabla \times \mathbf{H} = \epsilon \frac{\partial \mathbf{E}}{\partial t}$$

so that Equation 1.68 becomes

$$\nabla \times \nabla \times \mathbf{E} = -\mu\epsilon \frac{\partial^2 \mathbf{E}}{\partial t^2} \tag{1.69}$$

which using the vector identity

$$\nabla \times \nabla \times \mathbf{F} \equiv \nabla(\nabla \cdot \mathbf{F}) - \nabla^2 \mathbf{F}$$

can be written as

$$\nabla\nabla \cdot \mathbf{E} - \nabla^2 \mathbf{E} = -\mu\epsilon \frac{\partial^2 \mathbf{E}}{\partial t^2}$$

Since $\rho_v = 0$ and hence $\nabla \cdot \mathbf{E} = 0$, we obtain

$$\nabla^2 \mathbf{E} - \mu\epsilon \frac{\partial^2 \mathbf{E}}{\partial t^2} = \mathbf{0} \tag{1.70}$$

which is the time-dependent vector Helmholtz equation or simply wave equation. If we had started the derivation with Equation 1.2, we would obtain the wave equation for $\mathbf{H}$ as

$$\nabla^2 \mathbf{H} - \mu\epsilon \frac{\partial^2 \mathbf{H}}{\partial t^2} = \mathbf{0} \tag{1.71}$$

The wave travels at the speed

$$u = \frac{1}{\sqrt{\mu\epsilon}} \tag{1.72}$$

The wave repeats itself at a regular interval. The frequency f is defined as

$$f = \frac{\omega}{2\pi} \tag{1.73}$$

The EM waves obey all other usual wave phenomena of reflection and refraction when incident on interfaces between two different media. It should be noted that each of the vector equations in Equations 1.70 and 1.71 has three scalar components, so that altogether, we have six scalar equations for E_x, E_y, E_z, H_x, H_y, and H_z. Thus each component of the wave equation has the form

$$\nabla^2\Phi - \frac{1}{u^2}\frac{\partial^2\Phi}{\partial t^2} = 0 \tag{1.74}$$

which is the scalar wave equation. Replacing the time derivative in Equation 1.74 by $(j\omega)^2$ yields the scalar wave equation in phasor representation as

$$\nabla^2\Phi + k^2\Phi = 0 \tag{1.75}$$

where k is the wavenumber or the propagation constant (in rad/m), given by

$$k = \frac{\omega}{u} = \frac{2\pi f}{u} = \frac{2\pi}{\lambda} \tag{1.76}$$

and λ is the wavelength. It is the minimum distance over which the phase changes by 2π radians.

1.8 Diffusion Equation

We can derive the diffusion equation directly from Maxwell's equations. We recall from Equation 1.2 that

$$\nabla \times \mathbf{H} = \mathbf{J} + \mathbf{J}_d$$

where $\mathbf{J} = \sigma\mathbf{E}$ is the conduction current density and $\mathbf{J}_d = \frac{\partial \mathbf{D}}{\partial t}$ is the displacement current density. For $\frac{\sigma}{\omega\epsilon} \ll 1$, $\mathbf{J}_d$ is negligibly small compared with $\mathbf{J}$. Hence

$$\nabla \times \mathbf{H} = \mathbf{J} \tag{1.77}$$

Also taking curl both sides of Equation 1.1,

$$\nabla \times \nabla \times \mathbf{E} = \nabla(\nabla \cdot \mathbf{E}) - \nabla^2\mathbf{E} = -\mu\frac{\partial}{\partial t}(\nabla \times \mathbf{H}) \tag{1.78}$$

Using Equation 1.77 and the fact that $\nabla \cdot \mathbf{E} = 0$, we obtain

$$\nabla^2\mathbf{E} = \mu\frac{\partial \mathbf{J}}{\partial t} \tag{1.79}$$

Replacing $\mathbf{E}$ with $\mathbf{J}/\sigma$, Equation 1.79 now becomes

$$\nabla^2 \mathbf{J} = \mu\sigma \frac{\partial \mathbf{J}}{\partial t} \tag{1.80}$$

Assuming time harmonic field with time factor $e^{j\omega t}$,

$$\nabla^2 \mathbf{J}_s = j\omega\mu\sigma \mathbf{J}_s \tag{1.81}$$

1.9 Classification of EM Problems

EM problems are classified depending on the particular item of interest, which could be one of the following [4]:

1. The solution region of the problem
2. The nature of the equations describing the problem
3. The associated boundary conditions

In fact, the above three items define a given problem uniquely. It will soon become evident that these classifications are sometimes not independent of each other.

1.9.1 Classification of Solution Regions

In terms of the solution or problem domain, the problem could be an interior problem, also called an inner, closed, or bounded problem, or an exterior problem, also variously called an outer, open, or unbounded problem. Consider the solution region R with boundary S, as shown in Figure 1.4. If part or all of S is at infinity, R is exterior/open, otherwise R is interior/closed. For example, wave propagation in a waveguide is an interior problem, whereas wave propagation in free space (scattering of EM waves by raindrops, and radiation from a dipole antenna) are exterior problems. A problem can also be classified in terms of the electrical, constitutive properties (σ, μ ϵ) of the solution region. As mentioned earlier, the solution region could be linear (or nonlinear), homogeneous (or inhomogeneous), and isotropic (or anisotropic).

1.9.2 Classification of Differential Equations

EM problems are classified in terms of the differential equations describing them. The equations could be differential or integral or both. Most EM problems can be stated in terms of an operator equation

$$\mathcal{L}\Phi = g \tag{1.82}$$

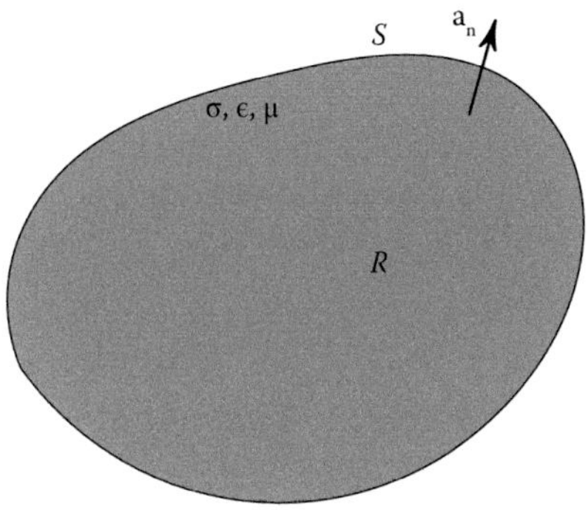

FIGURE 1.4
Solution region R with boundary S.

where $\mathcal{L}$ is an operator (differential, integral, or integro-differential), g is the known excitation or source, and Φ is the unknown function to be determined. A typical example is the electrostatic problem involving Poisson's equation

$$-\nabla^2 V = \frac{\rho_v}{\epsilon} \tag{1.83}$$

so that $\mathcal{L} = -\nabla^2$ is the Laplacian operator, $g = \frac{\rho_v}{\epsilon}$ is the source term, and $\Phi = V$ is the electric potential. In integral form, Poisson's equation is of the form

$$V = \int \frac{\rho_v}{4\pi\epsilon r} dv \tag{1.84}$$

so that

$$\mathcal{L} = \int \frac{dv}{4\pi r}, \qquad g = V, \qquad \Phi = \frac{\rho_v}{\epsilon} \tag{1.85}$$

In this section, we shall limit our discussion to differential equations. Solution of integral equations for arbitrary geometries becomes more amenable with numerical techniques than analytical methods. Only for a selected class of problems one could solve integral equations exactly. The reason for favoring numerical methods for integral equations is that the process of integration involves summation of several components. For the sum to be exact each individual summand need not to be exact. Any positive error in one element could be offset by a negative error in another element still resulting in an acceptable answer. An example for this is in computing the far-field radiation pattern of antennas. The exact current on the radiating structure need not to be known. A fairly good approximation would suffice. Most common numerical methods in EM are dealt in [4].

Most EM problems involve linear, second-order differential equations. In general, a second-order partial differential equation (PDE) is given by

$$a\frac{\partial^2 \Phi}{\partial x^2} + b\frac{\partial^2 \Phi}{\partial x\, \partial y} + c\frac{\partial^2 \Phi}{\partial y^2} + d\frac{\partial \Phi}{\partial x} + e\frac{\partial \Phi}{\partial y} + f\Phi = g \tag{1.86}$$

or simply

$$a\Phi_{xx} + b\Phi_{xy} + c\Phi_{yy} + d\Phi_x + e\Phi_y + f\Phi = g \tag{1.87}$$

The coefficients a, b, and c in general are functions of x and y; they may also depend on Φ itself, in which case the PDE is said to be nonlinear. A PDE in which $g(x, y) = 0$ in Equation 1.87 is termed homogeneous; it is inhomogeneous if $g(x, y) \neq 0$. Notice that Equation 1.87 has the same form as Equation 1.82, where $\mathcal{L}$ is now a differential operator given by

$$\mathcal{L} = a\frac{\partial^2}{\partial x^2} + b\frac{\partial^2}{\partial x\,\partial y} + c\frac{\partial^2}{\partial y^2} + d\frac{\partial}{\partial x} + e\frac{\partial}{\partial y} + f \tag{1.88}$$

A PDE in general can be associated with both boundary values and initial values. PDEs whose boundary conditions are specified are called steady-state equations. If only initial values are specified, they are called transient equations. Any linear second-order PDE can be classified as elliptic, hyperbolic, or parabolic depending on the coefficients a, b, and c. Equation 1.87 is said to be

$$\begin{array}{lll} \text{elliptic} & \text{if} & b^2 - 4ac < 0 \\ \text{hyperbolic} & \text{if} & b^2 - 4ac > 0 \\ \text{parabolic} & \text{if} & b^2 - 4ac = 0 \end{array} \tag{1.89}$$

The terms *hyperbolic, parabolic, and elliptic* are derived from the fact that the quadratic equation,

$$ax^2 + bxy + cy^2 + dx + ey + f = 0 \tag{1.90}$$

represents a hyperbola, parabola, or ellipse if $b^2 - 4ac$ is positive, zero, or negative, respectively. In each of these categories, there are PDEs that model certain physical phenomena. Such phenomena are not limited to EM but extend to almost all areas of science and engineering. Thus the mathematical model specified in Equation 1.87 arises in problems involving heat transfer, boundary-layer flow, vibrations, elasticity, electrostatic, wave propagation, and so on. Elliptic PDEs are associated with steady-state phenomena, that is, boundary-value problems. Typical examples of this type of PDE include Laplace's equation

$$\frac{\partial^2\Phi}{\partial x^2} + \frac{\partial^2\Phi}{\partial y^2} = 0 \tag{1.91}$$

and Poisson's equation

$$\frac{\partial^2\Phi}{\partial x^2} + \frac{\partial^2\Phi}{\partial y^2} = g(x, y) \tag{1.92}$$

where $a = c = 1$, $b = 0$. An elliptic PDE usually models an interior problem, and hence the solution region is usually closed or bounded as in Figure 1.5a. Hyperbolic PDEs arise in propagation problems. The solution region is usually open so that a solution advances outward indefinitely from initial conditions

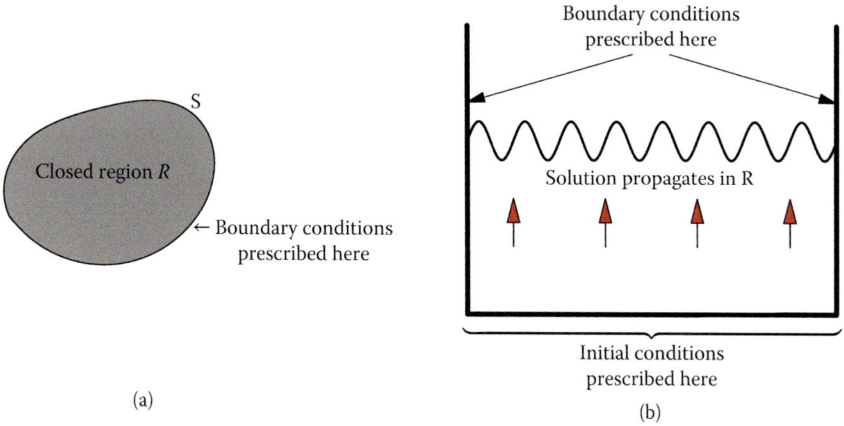

FIGURE 1.5
(a) Elliptic. (b) Parabolic or hyperbolic problem.

while always satisfying specified boundary conditions. A typical example of hyperbolic PDE is the wave equation in one dimension:

$$\frac{\partial^2 \Phi}{\partial x^2} = \frac{1}{u^2} \frac{\partial^2 \Phi}{\partial t^2} \tag{1.93}$$

where $a = u^2, \ b = 0, \ c = -1$. Notice that the wave equation in Equation 1.75 is not hyperbolic but elliptic since the time-dependence has been suppressed and the equation is merely the steady-state solution of Equation 1.74. Parabolic PDEs are generally associated with problems in which the quantity of interest varies slowly in comparison with the random motions which produce the variations. The most common parabolic PDE is the diffusion (or heat) equation in one dimension:

$$\frac{\partial^2 \Phi}{\partial x^2} = k \frac{\partial \Phi}{\partial t} \tag{1.94}$$

where $a = 1, \ b = 0 = c$. Like hyperbolic PDE, the solution region for parabolic PDE is usually open, as in Figure 1.5b. The initial and boundary conditions typically associated with parabolic equations resemble those for hyperbolic problems except that only one initial condition at $t = 0$ is necessary since Equation 1.94 is only first order in time. Also, parabolic and hyperbolic equations are solved using similar techniques, whereas elliptic equations require different techniques. Note that: (1) since the coefficients a, b, and c are in general functions of x and y, the classification of Equation 1.82 may change from point to point in the solution region, and (2) PDEs with more than two independent variables $(x, y, z, t, \cdots)$ may not fit as neatly into the classification above. A summary of our discussion so far in this section is shown in Table 1.2.

TABLE 1.2
Classification of Partial Differential Equations

Type	$b^2 - 4ac$	Example	Solution Region
Elliptic	< 0	Laplace's equation $\Phi_{xx} + \Phi_{yy} = 0$	Closed
Hyperbolic	> 0	Wave equation $u^2\Phi_{xx} = \Phi_{tt}$	Open
Parabolic	0	Diffusion equation $\Phi_{xx} = k\Phi_t = 0$	Open

The type of problem represented by Equation 1.82 is said to be deterministic since the quantity of interest can be determined directly. Another type of problem where the quantity is found indirectly is called nondetermistic or eigenvalue. The standard eigenproblem is of the form

$$\mathcal{L}\Phi = \lambda\Phi \tag{1.95}$$

where the source term in Equation 1.82 has been replaced by $\lambda\Phi$. A more general version is the generalized eigenproblem having the form

$$\mathcal{L}\Phi = \lambda\mathcal{M}\Phi \tag{1.96}$$

where $\mathcal{M}$, like $\mathcal{L}$, is a linear operator for EM problems. In Equations 1.95 and 1.96, only some particular values of λ called eigenvalues are permissible; associated with these values are the corresponding solutions Φ called eigenfunctions. Eigenproblems are usually encountered in vibration and waveguide problems where the eigenvalues λ correspond to physical quantities such as resonance and cutoff frequencies, respectively.

1.9.3 Classification of Boundary Conditions

Our problem consists of finding the unknown function Φ of a PDE. In addition to the fact that Φ satisfies Equation 1.82 within a prescribed solution region R, Φ must satisfy certain conditions on S, the boundary of R. Usually these boundary conditions are of the Dirichlet and Neumann types. Where a boundary has both, a mixed boundary condition is said to exist.

1. Dirichlet boundary condition:

$$\Phi(\mathbf{r}) = 0 \qquad \mathbf{r} \in S \tag{1.97}$$

2. Neumann boundary condition:

$$\frac{\partial\Phi(\mathbf{r})}{\partial n} = 0 \qquad \mathbf{r} \in S \tag{1.98}$$

that is, the normal derivative of Φ vanishes on S.

3. Mixed boundary condition:

$$\frac{\partial \Phi(\mathbf{r})}{\partial n} + h(\mathbf{r}) = 0 \qquad \mathbf{r} \in S \tag{1.99}$$

where $h(\mathbf{r})$ is a known function and $\frac{\partial \Phi(\mathbf{r})}{\partial n}$ is the directional derivative of Φ along the outward normal to the boundary S, that is

$$\frac{\partial \Phi(\mathbf{r})}{\partial n} = \nabla \Phi \cdot \mathbf{a}_n \tag{1.100}$$

where $\mathbf{a}_n$ is a unit normal directed out of R, as shown in Figure 1.4. Note that the Neumann boundary condition is a special case of the mixed condition with $h(\mathbf{r}) = 0$.

The conditions in Equations 1.97 through 1.100 are called homogeneous boundary conditions. The more general ones are the inhomogeneous.

1. Dirichlet boundary condition:

$$\Phi(\mathbf{r}) = p(\mathbf{r}) \qquad \mathbf{r} \in S \tag{1.101}$$

2. Neumann boundary condition:

$$\frac{\partial \Phi(\mathbf{r})}{\partial n} = q(\mathbf{r}) \qquad \mathbf{r} \in S \tag{1.102}$$

that is, the normal derivative of Φ vanishes on S.

3. Mixed boundary condition:

$$\frac{\partial \Phi(\mathbf{r})}{\partial n} + h(\mathbf{r}) = w(\mathbf{r}) \qquad \mathbf{r} \in S \tag{1.103}$$

where $h(\mathbf{r})$ is a known function and $\frac{\partial \Phi(\mathbf{r})}{\partial n}$ is the directional derivative of Φ along the outward normal to the boundary S, that is

$$\frac{\partial \Phi(\mathbf{r})}{\partial n} = \nabla \Phi \cdot \mathbf{a}_n \tag{1.104}$$

where $p(\mathbf{r}), q(\mathbf{r})$, and $w(\mathbf{r})$ are explicitly known functions on the boundary S.

For example, $\Phi(0) = 1$ is an inhomogeneous Dirichlet boundary condition, and the associated homogeneous counterpart is $\Phi(0) = 0$. Also, $\Phi'(1) = 2$ and $\Phi'(1) = 0$ are, respectively, inhomogeneous and homogeneous Neumann boundary conditions. In electrostatics, for example, if the value of electric potential is specified on S, we have Dirichlet boundary condition, whereas if the surface charge $\rho_s = D_n = \epsilon \frac{\partial V}{\partial n}$ is specified, the boundary condition is Neumann. The problem of finding a function Φ that is harmonic in a region is called *Dirichlet problem* (or *Neumann problem*) if Φ (or $\frac{\partial\Phi}{\partial n}$) is prescribed on the boundary of the region. It is worth observing that the term *homogeneous* has been used to mean different things. The solution region R could be homogeneous meaning that σ, μ, and ϵ are constant within R; the PDE could be homogeneous if $g = 0$ so that $\mathcal{L}\Phi = 0$; and the boundary conditions are homogeneous when $p(\mathbf{r}) = q(\mathbf{r}) = w(\mathbf{r}) = 0$.

References

[1] R. F. Harrington, *Time-Harmonic Electromagnetic Fields,* New York, Wiley-Interscience, IEEE Press, 2001.

[2] R. E. Collin, *Field Theory of Guided Waves*, New York: IEEE Press, 1991.

[3] C. A. Balanis, *Advanced Engineering Electromagnetics,* New York: John Wiley & Sons, 1989.

[4] M. N. O. Sadiku, *Numerical Techniques in Electromagnetics*, 3rd ed., Boca Raton: CRC Press, 2009.

Problems

1. Starting with the Maxwell's equations, show that

   (a) $\nabla \times \nabla \times \mathbf{H} + \mu\epsilon \frac{\partial^2 \mathbf{H}}{\partial t^2} = \nabla \times \mathbf{J}$

   (b) $\nabla \times \nabla \times \mathbf{E} + \mu\epsilon \frac{\partial^2 \mathbf{E}}{\partial t^2} = -\mu \frac{\partial \mathbf{J}}{\partial t}$

2. In a homogeneous, lossless, source-free medium,

$$\begin{aligned} \mathbf{E}_s &= E_0 e^{-j\beta z} \mathbf{a_x} \\ \mathbf{H}_s &= \frac{E_0}{\eta} e^{-j\beta z} \mathbf{a_y} \end{aligned}$$

   Find β and η for which $\mathbf{E}_s$ and $\mathbf{H}_s$ satisfy Maxwell's equations.

3. Show that a time-harmonic EM field in a conducting medium ($\sigma \gg \omega\epsilon$) satisfies the diffusion equation

$$\nabla^2 \mathbf{E}_s - j\omega\mu\sigma\mathbf{E}_s = \mathbf{0}$$

4. The electric field vector inside an air-filled rectangular waveguide is given by

$$\mathbf{E} = \sin\frac{\pi x}{a} e^{-j\beta z}\mathbf{a_x}$$

Find the associated magnetic field vector using Maxwell equations.

5. Consider a circular sheet of radius a, lying in the xy-plane described by $x^2 + y^2 \leq a^2$, $z = 0$. Suppose it carries a surface current $\mathbf{J}_s = J_0\mathbf{a_x}$. Find the magnetic vector potential at $(0, 0, z)$. Specialize your answer for $a \to \infty$.

6. Suppose the surface current $\mathbf{J}_s = J_0\mathbf{a_x}$ flows all over the xy-plane, find $\mathbf{A}$ using Equation 1.47 and determine the field vectors $\mathbf{E}$ and $\mathbf{H}$. Verify whether the boundary condition on $\mathbf{H}$ is satisfied.

7. Assuming that a given medium is lossy and the fields are time harmonic, state and prove uniqueness theorem for solution of wave equations. How do you justify your argument if the medium is lossless?

8. Consider the region of volume τ bounded by a closed surface S. Let $\mathbf{n}$ indicate unit outward normal on S. Suppose the two vector fields $\mathbf{A}$ and $\mathbf{B}$ are well-behaved inside τ. Show that

$$\iiint_{\tau} (\mathbf{A}\cdot\nabla\times\nabla\times\mathbf{B} - \mathbf{B}\cdot\nabla\times\nabla\times\mathbf{A})\,dv$$
$$= \oiint_{S} [\mathbf{B}\times(\nabla\times\mathbf{A}) - \mathbf{A}\times(\nabla\times\mathbf{B})]\cdot\mathbf{n}\,dS$$

9. In a region of space Ω, let $\mathbf{r}$, $\mathbf{r}'$ be the position vectors of points F and S, respectively and let $f(\mathbf{r})$ be a well-behaved scalar valued function. Show that

$$\iiint_{\Omega} f(\mathbf{r}')\nabla^2\left(\frac{1}{|\mathbf{r}-\mathbf{r}'|}\right)dv' + 4\pi f(\mathbf{r}) = 0$$

2

Separation of Variables

Religion is too serious a matter to be left to the priests.

– Anonymous

CONTENTS

Separation of variables (SoV) is the most universal and powerful analytical method for finding exact solutions to partial differential equations (PDEs). The key idea of SoV is to reduce the given PDE to several ordinary differential equations (ODEs), which can easily be solved. In this chapter, we begin by presenting the basic idea behind SoV—the conditions for complete separability. Then, we consider applying SOV to PDEs in rectangular, cylindrical, and spherical coordinates. In each coordinate system, we consider solutions to Laplace's equation (elliptic), heat equation (parabolic), and wave equation (hyperbolic) in 1D, 2D, and 3D.

2.1 Conditions for Complete Separability

The method of SoV is often attributed to Fourier [1] who solved the heat equation

$$\frac{\partial U}{\partial t} = \frac{\partial^2 U}{\partial x^2} \tag{2.1}$$

for heat distribution $U(x, t)$ in a metal rod, $0 < x < L$, looking for the product solution of the type

$$U(x, t) = X(x)T(t) \tag{2.2}$$

Since Fourier introduced SoV, it has become an indispensable tool in field theory, engineering, physics, chemistry, and biology. In general, for a PDE in Equation 2.1 with one dependent variable U and two independent variables x and t, the solutions of the form

$$U(x, t) = X(x)T(t) \qquad \text{and} \qquad U(x, t) = X(x) + T(t) \tag{2.3}$$

are known as product or additive solutions, respectively. Basically, applying SoV entails seeking a solution that breaks up into a product of functions, each of which involves only one of the variables. For example, if we are seeking a solution $\Phi(x, y, z, t)$ to some PDE, we require that it has the product form

$$\Phi(x, y, z, t) = X(x)Y(y)Z(z)T(t) \tag{2.4}$$

A solution of the form in Equation 2.4 is said to be separable in x, y, z, and t. For example, consider the following functions:

1. $x^2yz \sin 10t$
2. $x^2y^2 + \frac{1}{t}$
3. $(2x + y^2)z \cos 10t$

Among the above, (1) is completely separable, (2) is not separable, while (3) is separable only in z and t. To determine whether the method of separation of independent variables can be applied to a given physical problem, we must consider the PDE describing the problem, the shape of the solution region, and the boundary conditions—the three elements that uniquely define a problem. For example, to apply the method to a homogeneous problem such as $\mathcal{L}\Phi = 0$ involving two variables x and y (or ρ and φ, etc.), three things must be considered:

(i) The differential operator $\mathcal{L}$ must be separable; that is, it must be a function of $\Phi(x, y)$ such that

$$\frac{\mathcal{L}\{X(x)Y(y)\}}{X(x)Y(y)} \tag{2.5}$$

is a sum of a function of x only and a function of y only.

(ii) All initial and boundary conditions must be on constant-coordinate surfaces; that is, $x = \text{constant}$, $y = \text{constant}$.

(iii) The linear operators defining the boundary conditions at $x = \text{constant}$ (or $y = \text{constant}$) must involve no partial derivatives of Φ with respect to y (or x), and their coefficient must be independent of y (or x).

For example, the operator equation

$$\mathcal{L}\Phi = \frac{\partial^2 \Phi}{\partial x^2} + \frac{\partial^2 \Phi}{\partial x\, \partial y} + \frac{\partial^2 \Phi}{\partial y^2} \tag{2.6}$$

violates (i). If the solution region R is not a rectangle with sides parallel to the x and y axes, (ii) is violated. With a boundary condition $\Phi = 0$ on a part of $x = 0$ and $\frac{\partial \Phi}{\partial x} = 0$ on another part, (iii) is violated. With this preliminary discussion, we will now apply the method of SoV to PDEs in rectangular, circular cylindrical, and spherical coordinate systems. In each of these applications, we shall always take the following three major steps:

1. Separate the (independent) variables.
2. Find particular solutions of the separated equations, which satisfy some of the boundary conditions.
3. Combine these solutions to satisfy the remaining boundary conditions.

We begin the application of SoV by finding the product solution of the homogenous scalar wave equation

$$\nabla^2 \Phi - \frac{1}{c^2} \frac{\partial^2 \Phi}{\partial t^2} = 0 \tag{2.7}$$

Solution to Laplace's equation can be derived as a special case of the wave equation. Diffusion or heat equation can be handled in the same manner as we treat the wave equation. To solve Equation 2.7, it is expedient that we first separate the time dependence. We let

$$\Phi(\mathbf{r}, t) = U(\mathbf{r}) T(t) \tag{2.8}$$

Substituting this in Equation 2.7,

$$T \nabla^2 U - \frac{1}{c^2} U T'' = 0 \tag{2.9}$$

Dividing by UT gives

$$\frac{\nabla^2 U}{U} = \frac{T''}{c^2 T} \tag{2.10}$$

The left side is independent of t, while the right side independent of $\mathbf{r}$; the equality can be true only if each side is independent of both variables. If we let an arbitrary constant $-k^2$ be the common value of the two sides, Equation 2.10 reduces to

$$T'' + c^2k^2T = 0 \tag{2.11}$$

$$\nabla^2 U + k^2 U = 0 \tag{2.12}$$

Thus, we have been able to separate the space variable **r** from the time variable t. The arbitrary constant $-k^2$ introduced in the course of the SoV is called the separation constant. We shall see that in general the total number of independent separation constants in a given problem is one less than the number of independent variables involved. Equation 2.11 is an ODE with solution

$$T(t) = a_1 e^{jckt} + a_2 e^{-jckt} \tag{2.13a}$$

or

$$T(t) = b_1 \cos ckt + b_2 \sin ckt \tag{2.13b}$$

Since the time dependence does not change with a coordinate system, the time dependence expressed in Equation 2.13a is the same for all coordinate systems. Therefore, we shall henceforth restrict our effort to seeking a solution to Equation 2.12. Notice that if $k = 0$ then T either does not vary with time or is a linear function of time. Both of these two outcomes do not constitute wave motion, so we will not accept them as solution to wave equation. Thus, under the special case of $k = 0$, Equation 2.12 becomes Laplace's equation.

2.2 Rectangular Coordinates

To make things simpler, at the outset we shall first consider Laplace's equation in two dimensions. We will later consider the heat equation in one dimension and wave equation in three dimensions.

2.2.1 Laplace's Equation

Consider the Dirichlet problem of an infinitely long rectangular trough whose cross section is shown in Figure 2.1. For simplicity, let three of its sides be maintained at zero potential while the fourth side is at a fixed potential V_0. This is a boundary value problem. The PDE to be solved is

$$\frac{\partial^2 V}{\partial x^2} + \frac{\partial^2 V}{\partial y^2} = 0 \tag{2.14}$$

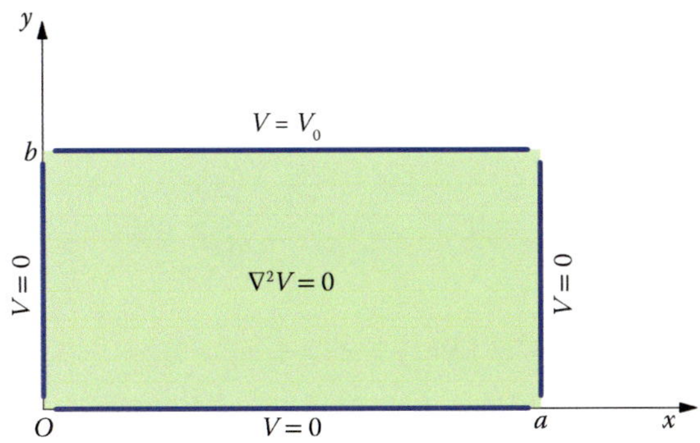

FIGURE 2.1
Cross-section of the rectangular conducting trough.

subject to the (Dirichlet) boundary conditions:

$$V(0, y) = 0 \tag{2.15a}$$
$$V(a, y) = 0 \tag{2.15b}$$
$$V(x, 0) = 0 \tag{2.15c}$$
$$V(x, b) = V_0 \tag{2.15d}$$

We let the solution be

$$V(x, y) = X(x)Y(y) \tag{2.16}$$

Substitute this in Equation 2.14 and divide by XY. This leads to

$$\frac{X''}{X} + \frac{Y''}{Y} = 0$$

or

$$\frac{X''}{X} = -\frac{Y''}{Y} = \lambda \tag{2.17}$$

where λ is the separation constant. Thus, the separated equations are

$$X'' - \lambda X = 0 \tag{2.18}$$
$$Y'' + \lambda Y = 0 \tag{2.19}$$

To solve the ODEs 2.18 and 2.19, we must impose the boundary conditions in Equation 2.15. However, these boundary conditions must be transformed so that they can be applied directly to the separated equations. Since $V = XY$,

$$V(0, y) = 0 \rightarrow X(0) = 0 \tag{2.20a}$$
$$V(a, y) = 0 \rightarrow X(a) = 0 \tag{2.20b}$$
$$V(x, 0) = 0 \rightarrow Y(0) = 0 \tag{2.20c}$$
$$V(x, b) = V_0 \rightarrow X(x)Y(b) = V_0 \tag{2.20d}$$

Notice that only the homogeneous conditions are separable. To solve Equation 2.18, we distinguish the three possible cases: $\lambda = 0$, $\lambda > 0$, and $\lambda < 0$.

Case 1: If $\lambda = 0$, Equation 2.18 reduces to

$$X'' = 0 \qquad \text{or} \qquad \frac{d^2X}{dx^2} = 0 \tag{2.21}$$

which has the solution

$$X(x) = a_1 x + a_2 \tag{2.22}$$

where a_1 and a_2 are constants. Imposing the conditions in Equations 2.20a and 2.20b,

$$\begin{aligned} X(0) = 0 \quad &\rightarrow \quad a_2 = 0 \\ X(a) = 0 \quad &\rightarrow \quad a_1 = 0 \end{aligned}$$

Hence, $X(x) = 0$, a trivial solution. This renders case $\lambda = 0$ unacceptable.

Case 2: If $\lambda > 0$, say $\lambda = \alpha^2$, Equation 2.18 becomes

$$X'' - \alpha^2 X = 0 \tag{2.23}$$

The corresponding auxiliary equation is $m^2 - \alpha^2 = 0$ and the roots are $m = \pm\alpha$. Hence, the general solution is

$$X = b_1 e^{-\alpha x} + b_2 e^{\alpha x} \tag{2.24}$$

or

$$X = b_3 \sinh \alpha x + b_4 \cosh \alpha x \tag{2.25}$$

The boundary conditions are applied to determine constants b_3 and b_4.

$$\begin{aligned} X(0) = 0 \quad &\rightarrow \quad b_4 = 0 \\ X(a) = 0 \quad &\rightarrow \quad b_3 = 0 \end{aligned}$$

since $\sinh \alpha x$ is never zero for $\alpha > 0$. Hence, $X(x) = 0$, a trivial solution, and we conclude that case $\lambda > 0$ is also not valid.

Case 3: If $\lambda < 0$, say $\lambda = -\beta^2$, Equation 2.18 becomes

$$X'' + \beta^2 X = 0 \tag{2.26}$$

with the auxiliary equation $m^2 + \beta^2 = 0$ or $m = \pm j\beta$. The solution to Equation 2.26 is

$$X = A_1 e^{j\beta x} + A_2 e^{-j\beta x} \tag{2.27a}$$

or

$$X = B_1 \sin \beta x + B_2 \cos \beta x \tag{2.27b}$$

Again,

$$X(0) = 0 \quad \rightarrow \quad B_2 = 0$$
$$X(a) = 0 \quad \rightarrow \quad \sin \beta a = 0 = \sin n\pi$$

or

$$\beta = \frac{n\pi}{a} \qquad n = 1, 2, 3, \cdots \tag{2.28}$$

since B_1 cannot vanish for nontrivial solutions, whereas $\sin \beta a$ can vanish without its argument being zero. Thus, we have found an infinite set of discrete values of λ for which Equation 2.18 has nontrivial solutions; that is,

$$\lambda = -\beta^2 = -\frac{n^2\pi^2}{a^2} \qquad n = 1, 2, 3, \cdots \tag{2.29}$$

These are the eigenvalues of the problem and the corresponding eigenfunctions are

$$X_n(x) = \sin \beta x = \sin \frac{n\pi x}{a} \tag{2.30}$$

From Equation 2.29, note that it is not necessary to include negative values of n since they lead to the same set of eigenvalues. Also, we exclude $n = 0$ since it yields the trivial solution $X = 0$ as shown under case 1 when $\lambda = 0$. Having determined λ, we now proceed to solve Equation 2.19 in order to find $Y_n(y)$ corresponding to $X_n(x)$. That is, we solve

$$Y'' - \beta^2 Y = 0 \tag{2.31}$$

This is similar to Equation 2.23, whose solution is of the form Equation 2.25. Hence, the solution to Equation 2.31 has the form

$$Y_n(y) = a_n \sinh \frac{n\pi y}{a} + b_n \cosh \frac{n\pi y}{a} \tag{2.32}$$

the boundary condition in Equation 2.20c,

$$Y(0) = 0 \quad \rightarrow \quad b_n = 0$$

so that,

$$Y_n(y) = a_n \sinh \frac{n\pi y}{a} \tag{2.33}$$

Substituting Equations 2.30 and 2.33 into Equation 2.16, we obtain

$$V_n(x, y) = X_n(x)Y_n(y) = a_n \sin \frac{n\pi x}{a} \sinh \frac{n\pi y}{a} \tag{2.34}$$

which satisfies Equation 2.14 and the three homogeneous boundary conditions in Equations 2.15a, 2.15b, and 2.15c. By the superposition principle, a linear combination of the solutions V_n, each with different values of n and arbitrary coefficient a_n, is also a solution of Equation 2.14. Thus, we may represent the solution V of Equation 2.14 as an infinite series in the functions V_n; that is,

$$V(x,y) = \sum_{n=1}^{\infty} a_n \sin\frac{n\pi x}{a} \sinh\frac{n\pi y}{a} \tag{2.35}$$

We now determine the coefficients a_n by imposing the inhomogeneous boundary condition of Equation 2.15d in Equation 2.35. We get

$$V(x,b) = \sum_{n=1}^{\infty} a_n \sin\frac{n\pi x}{a} \sinh\frac{n\pi b}{a} \tag{2.36}$$

which is Fourier sine expansion. Hence,

$$a_n \sinh\frac{n\pi b}{a} = \frac{2}{a}\int_0^a V_0 \sin\frac{n\pi x}{a} \mathrm{d}x = \frac{2V_0}{n\pi}(1-\cos n\pi)$$

or

$$a_n = \begin{cases} \frac{4V_0}{n\pi}\left(\frac{1}{\sinh\frac{n\pi b}{a}}\right) & \text{if } n \text{ is odd} \\ 0 & \text{if } n \text{ is even} \end{cases} \tag{2.37}$$

Substitution of Equation 2.37 into Equation 2.35 gives the complete solution as

$$V(x,y) = \frac{4V_0}{\pi} \sum_{n=\text{odd}} \left(\frac{\sin\frac{n\pi x}{a} \sinh\frac{n\pi y}{a}}{n \sinh\frac{n\pi b}{a}} \right) \tag{2.38}$$

By replacing n by $2k-1$, Equation 2.38 may be written as

$$V(x,y) = \frac{4V_0}{\pi} \sum_{k=1}^{\infty} \left(\frac{\sin\frac{(2k-1)\pi x}{a} \sinh\frac{(2k-1)\pi y}{a}}{(2k-1)\sinh\frac{(2k-1)\pi b}{a}} \right) \tag{2.39}$$

The problem we have solved has only one inhomogeneous boundary condition. We now reconsider the problem in Figure 2.1, but with four inhomogeneous boundary conditions as in Figure 2.2 (top row). The problem can be stated as solving Laplace's equation

$$\frac{\partial^2 V}{\partial x^2} + \frac{\partial^2 V}{\partial y^2} = 0 \tag{2.40}$$

subject to the following inhomogeneous Dirichlet boundary conditions

$$V(x,0) = V_1 \tag{2.41a}$$
$$V(x,b) = V_3 \tag{2.41b}$$
$$V(0,y) = V_4 \tag{2.41c}$$
$$V(a,y) = V_2 \tag{2.41d}$$

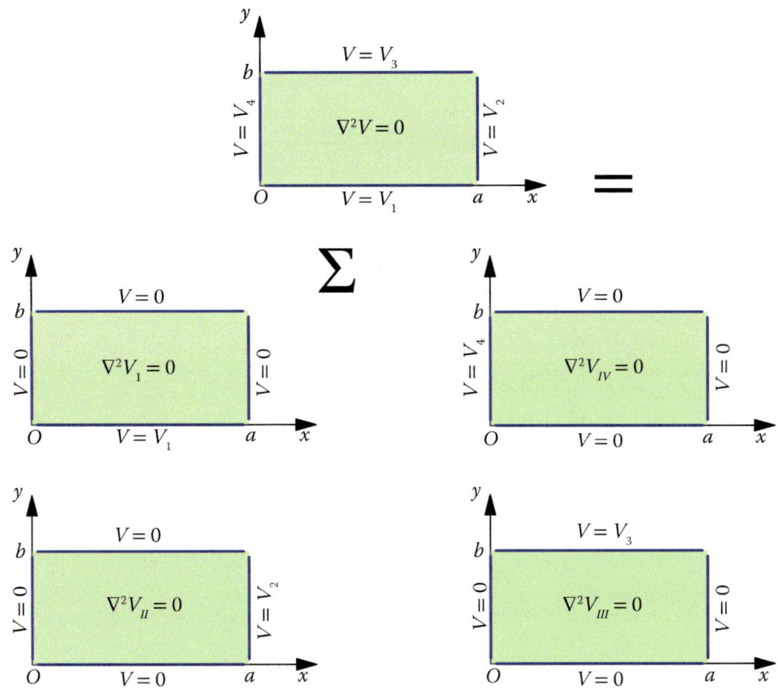

FIGURE 2.2
Applying the principle of superposition reduces the problem involving four different potentials on four walls to a combination of four subproblems with only one potential on one wall.

Since Laplace's equation is a linear homogeneous equation, the problem can be simplified by applying the superposition principle. If we let

$$V = V_I + V_{II} + V_{III} + V_{IV} \tag{2.42}$$

We may reduce the problem to four simpler problems, each of which is associated with the (say,) first of the inhomogeneous conditions. The reduced, simpler problems are illustrated in Figure 2.2 and stated as follows.

$$\frac{\partial^2 V_I}{\partial x^2} + \frac{\partial^2 V_I}{\partial y^2} = 0 \tag{2.43}$$

subject to the following inhomogeneous Dirichlet boundary conditions

$$V_I(x, 0) = V_1 \tag{2.44a}$$

$$V_I(x, b) = 0 \tag{2.44b}$$

$$V_I(0, y) = 0 \tag{2.44c}$$

$$V_I(a, y) = 0 \tag{2.44d}$$

$$\frac{\partial^2 V_{II}}{\partial x^2} + \frac{\partial^2 V_{II}}{\partial y^2} = 0 \tag{2.45}$$

subject to

$$V_{II}(x,0) = 0 \tag{2.46a}$$
$$V_{II}(x,b) = 0 \tag{2.46b}$$
$$V_{II}(0,y) = 0 \tag{2.46c}$$
$$V_{II}(a,y) = V_2 \tag{2.46d}$$

$$\frac{\partial^2 V_{III}}{\partial x^2} + \frac{\partial^2 V_{III}}{\partial y^2} = 0 \tag{2.47}$$

subject to

$$V_{III}(x,0) = 0 \tag{2.48a}$$
$$V_{III}(x,b) = V_3 \tag{2.48b}$$
$$V_{III}(0,y) = 0 \tag{2.48c}$$
$$V_{III}(a,y) = 0 \tag{2.48d}$$

$$\frac{\partial^2 V_{IV}}{\partial x^2} + \frac{\partial^2 V_{IV}}{\partial y^2} = 0 \tag{2.49}$$

subject to

$$V_{IV}(x,0) = 0 \tag{2.50a}$$
$$V_{IV}(x,b) = 0 \tag{2.50b}$$
$$V_{IV}(0,y) = V_4 \tag{2.50c}$$
$$V_{IV}(a,y) = 0 \tag{2.50d}$$

It is obvious that the reduced problem in Equations 2.47 and 2.48 with solution V_{III} is the same as that in Figure 2.1. The other three reduced problems are quite similar. Hence, the solutions V_I, V_{II}, and V_{IV} can be obtained by taking the same steps or by proper exchange of variables in Equation 2.38. Thus,

$$V_I(x,y) = \frac{4V_1}{\pi} \sum_{k=1}^{\infty} \left(\frac{\sin \frac{(2k-1)\pi x}{a} \sinh \frac{(2k-1)\pi(b-y)}{a}}{(2k-1)\sinh \frac{(2k-1)\pi b}{a}} \right) \tag{2.51}$$

$$V_{II}(x,y) = \frac{4V_2}{\pi} \sum_{k=1}^{\infty} \left(\frac{\sin \frac{(2k-1)\pi x}{b} \sinh \frac{(2k-1)\pi y}{b}}{(2k-1)\sinh \frac{(2k-1)\pi a}{b}} \right) \tag{2.52}$$

$$V_{III}(x,y) = \frac{4V_3}{\pi} \sum_{k=1}^{\infty} \left(\frac{\sin \frac{(2k-1)\pi x}{a} \sinh \frac{(2k-1)\pi y}{a}}{(2k-1)\sinh \frac{(2k-1)\pi b}{a}} \right) \tag{2.53}$$

$$V_{IV}(x,y) = \frac{4V_4}{\pi} \sum_{k=1}^{\infty} \left(\frac{\sin \frac{(2k-1)\pi(a-x)}{b} \sinh \frac{(2k-1)\pi y}{b}}{(2k-1)\sinh \frac{(2k-1)\pi b}{b}} \right) \tag{2.54}$$

We obtain the complete solution by substituting Equations 2.51 through 2.54 in Equation 2.42.

2.2.2 Heat Equation

To be concrete, suppose we are seeking a product solution of the heat or diffusion equation

$$\Phi_t = k\Phi_{xx} \qquad 0 < x < 1, \quad t > 0 \tag{2.55}$$

subject to the boundary conditions

$$\Phi(0, t) = 0 = \Phi(1, t) \qquad t > 0 \tag{2.56}$$

and the initial condition

$$\Phi(x, 0) = 5\sin 2\pi x \qquad 0 < x < 1 \tag{2.57}$$

The solution region for the problem is illustrated in Figure 2.3. To find the solution, we let

$$\Phi(x, t) = X(x)T(t) \tag{2.58}$$

Substitute this in Equation 2.55 and divide by kXT to obtain

$$\frac{T'}{kT} = \frac{X''}{X} = \lambda \tag{2.59}$$

where λ is the separation constant. Thus, the separated equations are

$$X'' - \lambda X = 0 \tag{2.60}$$
$$T' - k\lambda T = 0 \tag{2.61}$$

As usual, in order for the solution of Equation 2.60 to satisfy Equation 2.56, we must choose $\lambda = -n^2\pi^2$, $n = 1, 2, 3, \cdots$ so that

$$X_n(x) = \sin n\pi x \tag{2.62}$$

Equation 2.61 becomes

$$T' + kn^2\pi^2 T = 0$$

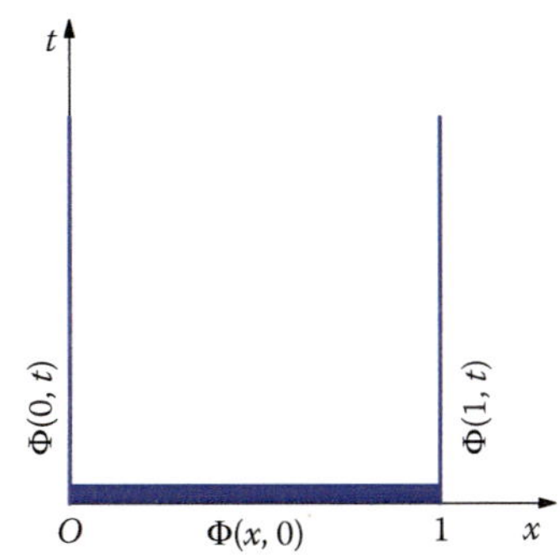

FIGURE 2.3
Solution region for the heat equation.

which has the solution

$$T_n(t) = a_n \mathrm{e}^{-kn^2\pi^2 t} \tag{2.63}$$

Substituting Equations 2.62 and 2.63 into Equation 2.58,

$$\Phi_n(x,t) = a_n \sin n\pi x \mathrm{e}^{-kn^2\pi^2 t}$$

where the coefficients a_n are to be determined from the initial condition in Equation 2.57. The complete solution is a linear combination of Φ_n; that is,

$$\Phi(x,t) = \sum_{n=1}^{\infty} a_n \sin n\pi x \mathrm{e}^{-kn^2\pi^2 t}$$

This satisfies Equation 2.57 if

$$\Phi(x,0) = \sum_{n=1}^{\infty} a_n \sin n\pi x = 5 \sin 2\pi x \tag{2.64}$$

The coefficients a_n are determined as

$$a_n = \frac{2}{L}\int_0^1 5 \sin 2\pi x \sin n\pi x dx = \begin{cases} 5 & n = 2 \\ 0 & n \neq 2 \end{cases} \tag{2.65}$$

where $L = 1$ is the length of the spatial interval. Alternatively, by comparing the middle term in Equation 2.64 with the last term, the two are equal only when $n = 2$, $a_n = 5$, otherwise $a_n = 0$. Hence, the solution of the diffusion problem becomes

$$\Phi(x,t) = 5 \sin 2\pi t \mathrm{e}^{-4k\pi^2 t} \tag{2.66}$$

2.2.3 Wave Equation

Let us now attempt a product solution of the Helmholtz wave equation

$$\nabla^2 U + k^2 U = 0 \tag{2.67}$$

In rectangular coordinates, Equation 2.67 becomes

$$\frac{\partial^2 U}{\partial x^2} + \frac{\partial^2 U}{\partial y^2} + \frac{\partial^2 U}{\partial z^2} + k^2 U = 0 \tag{2.68}$$

We let

$$U(x,y,z) = X(x)Y(y)Z(z) \tag{2.69}$$

Substituting Equation 2.69 into Equation 2.68 and dividing by XYZ, we obtain

$$\frac{X''}{X} + \frac{Y''}{Y} + \frac{Z''}{Z} + k^2 = 0 \tag{2.70}$$

Each term must be equal to a constant since each term depends only on the corresponding variable; X on x, Y on y, and so on. We conclude that

$$\frac{X''}{X} = -k_x^2, \quad \frac{Y''}{Y} = -k_y^2, \quad \frac{Z''}{Z} = -k_z^2 \tag{2.71}$$

so that Equation 2.70 reduces to

$$k_x^2 + k_y^2 + k_z^2 = k^2 \tag{2.72}$$

Notice that we have introduced three separation constants k_x, k_y, k_z as we came across three variables x, y, and z. But Equation 2.72 poses a constraint, so that only two of the separation constants are independent. As mentioned earlier, the number of independent separation constants is generally one less than the number of independent variables involved. The ODEs in Equation 2.71 have solutions

$$X = A_1 e^{jk_x x} + A_2 e^{-jk_x x} \tag{2.73a}$$

or

$$X = B_1 \sin k_x x + B_2 \cos k_x x \tag{2.73b}$$

$$Y = A_3 e^{jk_y y} + A_4 e^{-jk_y y} \tag{2.74a}$$

or

$$Y = B_3 \sin k_y y + B_4 \cos k_y y \tag{2.74b}$$

$$Z = A_5 e^{jk_z z} + A_6 e^{-jk_z z} \tag{2.75a}$$

or

$$Z = B_5 \sin k_z z + B_6 \cos k_z z \tag{2.75b}$$

Various combinations of X, Y, and Z will satisfy Equation 2.67. The actual combination selected will depend on the boundary and initial conditions. Since we are not given some specific boundary and initial conditions, suppose we choose

$$X = A_1 e^{jk_x x}, \qquad Y = A_3 e^{jk_y y}, \qquad Z = A_5 e^{jk_z z} \tag{2.76}$$

Combining the arbitrary constants into A, the product solution is

$$U(x, y, z) = A e^{jk_x x + jk_y y + jk_z z} \tag{2.77}$$

or

$$U(\mathbf{r}) = A e^{j\mathbf{k}\cdot\mathbf{r}} \tag{2.78}$$

In addition to spatial variation, if we also assume the $e^{i\omega t}$ time dependence (see Equation 2.13a) we get,

$$U(x, y, z, t) = A e^{j\mathbf{k}\cdot\mathbf{r} + j\omega t} \tag{2.79}$$

where ω is the angular frequency of the wave and k controls the choice of separation constants according to Equation 2.72. The solution in Equation 2.79 represents a plane wave of amplitude A propagating in the direction of the wave vector $\mathbf{k} = k_x \mathbf{a}_x + k_y \mathbf{a}_y + k_z \mathbf{a}_z$ with velocity c.

2.3 Cylindrical Coordinates

A problem having cylindrical symmetry is best solved in a cylindrical coordinate system, where the coordinate variables (ρ, φ, z) are described in Figure 2.4 with ranges of values

$$0 < \rho < \infty, \qquad 0 < \varphi < 2\pi, \qquad -\infty < z < \infty \tag{2.80}$$

As we did in the previous section, we shall first solve Laplace's equation. Then we shall take the heat conduction equation and wave equation.

2.3.1 Laplace's Equation

We will first consider a simple case of a potential function of two cylindrical variables, that is, $V = V(\rho, \varphi)$ and later proceed to the more general case of $V = V(\rho, \varphi, z)$.

Case A: $V = V(\rho, \varphi)$ Consider an infinitely long conducting cylinder of radius a with cross section shown in Figure 2.5. Assume that the upper half of the cylinder is maintained at potential V_0, while the lower half is maintained at potential $-V_0$. This is a Laplacian problem in two dimension. Hence, we solve for $V = V(\rho, \varphi)$ satisfying the Laplace's equation

$$\nabla^2 V = \frac{1}{\rho}\frac{\partial}{\partial \rho}\left(\rho \frac{\partial V}{\partial \rho}\right) + \frac{1}{\rho^2}\frac{\partial^2 V}{\partial \varphi^2} = 0 \tag{2.81}$$

Subject to the inhomogeneous Dirichlet boundary conditions

$$V(a, \varphi) = \begin{cases} V_0 & 0 < \varphi < \pi \\ -V_0 & \pi < \varphi < 2\pi \end{cases} \tag{2.82}$$

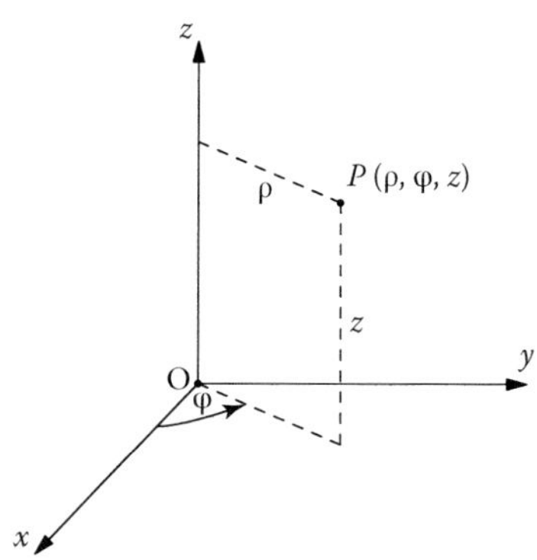

FIGURE 2.4
Coordinate relations in a cylindrical system.

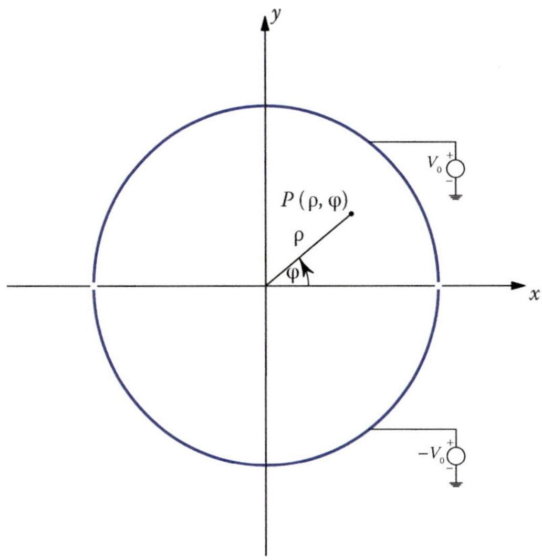

FIGURE 2.5
A two-dimensional Laplacian problem in cylindrical coordinates.

We let

$$V(a, \varphi) = R(\rho)F(\varphi) \tag{2.83}$$

Substituting Equation 2.83 into Equation 2.81 and dividing through by $\dfrac{RF}{\rho^2}$ result in

$$\frac{\rho}{R}\frac{\mathrm{d}}{\mathrm{d}\rho}\left(\rho\frac{\mathrm{d}R}{\mathrm{d}\rho}\right) + \frac{1}{F}\frac{\mathrm{d}^2F}{\mathrm{d}\varphi^2} = 0$$

or

$$\frac{\rho^2}{R}\frac{\mathrm{d}^2R}{\mathrm{d}\rho^2} + \frac{\rho}{R}\frac{\mathrm{d}R}{\mathrm{d}\rho} = -\frac{1}{F}\frac{\mathrm{d}^2F}{\mathrm{d}\varphi^2} = \lambda^2 \tag{2.84}$$

where λ is the separation constant. Thus, the separated equations are

$$F'' + \lambda^2 F = 0 \tag{2.85}$$

$$\rho^2 R'' + \rho R' - \lambda^2 R = 0 \tag{2.86}$$

It is evident that Equation 2.85 has the general solution of the form

$$F(\varphi) = c_1 \cos\lambda\varphi + c_2 \sin\lambda\varphi \tag{2.87}$$

From the boundary conditions of Equation 2.82, we notice that $F(\varphi)$ must be periodic, odd function. Thus, $c_1 = 0$. Also it is rather apt to take $\lambda = n$ a positive integer, and hence Equation 2.87 becomes

$$F_n(\varphi) = c_2 \sin n\varphi \tag{2.88}$$

Equation 2.86, known as the *Cauchy–Euler equation*, can be solved by making a substitution $\rho = e^u$ and reducing it to an equation with constant coefficients. This is done by writing

$$\begin{aligned}\frac{dR}{d\rho} &= \frac{dR}{du}\frac{du}{d\rho} = e^{-u}\frac{dR}{du}\\ \frac{d^2R}{d\rho^2} &= e^{-u}\frac{d}{du}\left(e^{-u}\frac{dR}{du}\right)\\ &= e^{-2u}\left(\frac{d^2R}{du^2} - \frac{dR}{du}\right)\end{aligned}$$

With these substitutions including $\lambda = n$, Equation 2.86 becomes

$$\frac{d^2R}{du^2} - n^2R = 0$$

whose solution for a specific value of n

$$R_n(u) = A_n e^{nu} + B_n e^{-nu}$$

leads us to write

$$R_n(\rho) = A_n\rho^n + B_n\rho^{-n} \tag{2.89}$$

Note that case $n = 0$ is excluded; if $n = 0$, we obtain the solution in the form $R_n(\rho) = c_3 \ln\rho + c_4$, which is not finite at $\rho = 0$. For the problem of a coaxial cable, $a < \rho < b$, $\rho \neq 0$ so that case $n = 0$ in the only solution. However, for the problem at hand, $n = 0$ is not acceptable. Substitution of Equations 2.88 and 2.89 into Equation 2.83 yields

$$V_n(\rho, \varphi) = \sin n\phi \left(A_n\rho^n + B_n\rho^{-n}\right) \tag{2.90}$$

where A_n and B_n are constants to be determined. As usual, it is possible by the superposition principle to form a complete series solution

$$V_n(\rho, \varphi) = \sum_{n=1}^{\infty} \left(A_n\rho^n + B_n\rho^{-n}\right) \sin n\varphi \tag{2.91}$$

For $\rho < a$, inside the cylinder, V must be finite as $\rho \to 0$ so that $B_n = 0$. At $\rho = a$,

$$V_n(a, \varphi) = \sum_{n=1}^{\infty} A_n a^n \sin n\varphi = \begin{cases} V_0 & 0 < \varphi < \pi \\ -V_0 & \pi < \varphi < 2\pi \end{cases} \tag{2.92}$$

Multiplying both sides by $\sin m\varphi$ and integrating over $0 < \varphi < 2\pi$, we get

$$\int_0^{\pi} V_0 \sin m\varphi d\varphi - \int_{\pi}^{2\pi} V_0 \sin m\varphi d\varphi = \sum_{n=1}^{\infty} A_n a^n \int_0^{2\pi} \sin m\varphi \sin n\varphi d\varphi$$

All terms in the right-hand side vanish except when $m = n$. Hence,

$$\frac{2V_0}{n}(1 - \cos n\pi) = A_n a^n \int_0^{2\pi} \sin^2 \varphi d\varphi = \pi A_n a^n$$

or

$$A_n = \begin{cases} \frac{4V_0}{n\pi a^n} & n = \text{odd} \\ 0 & n = \text{even} \end{cases} \tag{2.93}$$

Thus,

$$V(\rho, \varphi) = \frac{4V_0}{\pi} \sum_{n=\text{odd}} \frac{\rho^n \sin n\varphi}{na^n}, \qquad \rho < a \tag{2.94}$$

For $\rho > a$, outside the cylinder, V must be finite as $\rho \to \infty$ so that $A_n = u$ in Equation 2.91. By imposing the boundary condition in Equation 2.82 and following the same steps as for case $\rho < a$, we obtain

$$B_n = \begin{cases} \frac{4V_0 a^n}{n\pi} & n = \text{odd} \\ 0 & n = \text{even} \end{cases} \tag{2.95}$$

Hence,

$$V(\rho, \varphi) = \frac{4V_0}{\pi} \sum_{n=\text{odd}} \frac{a^n \sin n\varphi}{n\rho^n}, \qquad \rho > a \tag{2.96}$$

Case B: $V = V(\rho, \varphi, z)$ Laplace's equation in three dimensions in cylindrical coordinates is

$$\nabla^2 V = \frac{1}{\rho}\frac{\partial}{\partial \rho}\left(\rho \frac{\partial V}{\partial \rho}\right) + \frac{1}{\rho^2}\frac{\partial^2 V}{\partial \varphi^2} + \frac{\partial^2 V}{\partial z^2} = 0 \tag{2.97}$$

Assume

$$V(\rho, \varphi, z) = R(\rho)F(\varphi)Z(z) \tag{2.98}$$

Substituting Equation 2.97 into Equation 2.98

$$\frac{FZ}{\rho}\frac{\partial}{\partial \rho}\left(\rho \frac{\partial R}{\partial \rho}\right) + \frac{RZ}{\rho^2}\frac{\partial^2 F}{\partial \varphi^2} + RF\frac{\partial^2 Z}{\partial z^2} = 0$$

Divide through by RFZ, we get

$$\frac{1}{R\rho}\frac{\partial}{\partial \rho}\left(\rho \frac{\partial R}{\partial \rho}\right) + \frac{1}{F\rho^2}\frac{\partial^2 F}{\partial \varphi^2} + \frac{1}{Z}\frac{\partial^2 Z}{\partial z^2} = 0 \tag{2.99}$$

The Z term is isolated. We set it equal to k^2 and the solution is of the form

$$Z(z) = c_1 \cosh kz + c_2 \sinh kz \tag{2.100a}$$

or

$$Z(z) = Ae^{kz} + Be^{-kz} \tag{2.100b}$$

Equation 2.99 now becomes

$$\frac{\rho}{R}\frac{\partial}{\partial\rho}\left(\rho\frac{\partial R}{\partial\rho}\right)+\frac{1}{F}\frac{\partial^2 F}{\partial\varphi^2}+(k\rho)^2=0 \tag{2.101}$$

The F term is now isolated. Anticipating sinusoidal solutions, we set the F term equal to $-n^2$ and obtain the solution as

$$F(\varphi)=c_3\cos n\varphi+c_4\sin n\varphi \tag{2.102}$$

Continuity in V requires that n be a non-negative integer. Equation 2.101 now becomes

$$\rho\frac{\partial}{\partial\rho}\left(\rho\frac{\partial R}{\partial\rho}\right)+\left[(k\rho)^2-n^2\right]R=0$$

or

$$\rho^2\frac{d^2R}{d\rho^2}+\rho\frac{dR}{d\rho}+\left[(k\rho)^2-n^2\right]R=0 \tag{2.103}$$

To solve Equation 2.103, let $x=k\rho$ (not the x-coordinate, of course!) so that

$$\frac{dR}{d\rho}=\frac{dR}{dx}\frac{dx}{d\rho}=k\frac{dR}{dx}$$

Similarly,

$$\frac{d^2R}{d\rho^2}=k\frac{d}{dx}\frac{dR}{dx}\frac{dx}{d\rho}=k^2\frac{dR^2}{dx^2}$$

Substituting these in Equation 2.103, we obtain

$$x^2\frac{d^2R}{dx^2}+x\frac{dR}{dx}+\left[x^2-n^2\right]R=0$$

or

$$xR''+xR'+\left[x^2-n^2\right]R=0 \tag{2.104}$$

This is known as the Bessel's differential equation. It has a general solution of the form

$$R(x)=c_5J_n(x)+c_6Y_n(x) \tag{2.105}$$

where $J_n(x)$ and $Y_n(x)$ are respectively *Bessel functions* of the first and second kinds of order n and real argument x. Y_n is also called *Neumann function*. Substituting Equations 2.100, 2.102, and 2.105 into Equation 2.98 yields the complete general solution of Laplace's equation as

$$V(\rho,\varphi,z)=\sum_{n=0}^{\infty}\Big\{\left[A_nJ_n(k\rho)+B_nY_n(k\rho)\right]\left[C_n\sin n\varphi+D_n\cos n\varphi\right]\left[E_n\sinh kz+F_n\cosh kz\right]\Big\} \tag{2.106}$$

The expansion coefficients $A_n \cdots F_n$ are determined by imposing the boundary conditions. Since we are going to deal with Bessel function again, it is expedient that we spend some time now on it. If x in Equation 2.104 is imaginary so that we replace x by jx, and R' by $-jR'$, the equation becomes

$$xR'' + xR' - \left[x^2 + n^2\right] R = 0 \tag{2.107}$$

which is called the *modified Bessel's differential equation*. It has a general solution

$$R(x) = c_7 I_n(x) + c_8 K_n(x) \tag{2.108}$$

where $I_n(x)$ and $K_n(x)$ are respectively called the modified Bessel functions of the first and second kind of order n. For small values of x, Figures 2.6 through 2.9 show the plots of some typical Bessel functions (or cylindrical functions) $J_n(x)$, $Y_n(x)$, $I_n(x)$, and $K_n(x)$.

To obtain the Bessel functions from Equations 2.104 and 2.107, the method of Frobenius is applied. A detailed discussion is found in Myint-U [2]. For the Bessel function of the first kind,

$$J_n(x) = \sum_{m=0}^{\infty} \frac{(-1)^m (x/2)^{n+2m}}{m!\,\Gamma(n+m+1)} \tag{2.109}$$

where $\Gamma(k+1) = k!$ is the Gamma function. This is the most useful of all Bessel functions. Some of the important properties and identities are listed in Table 2.1.

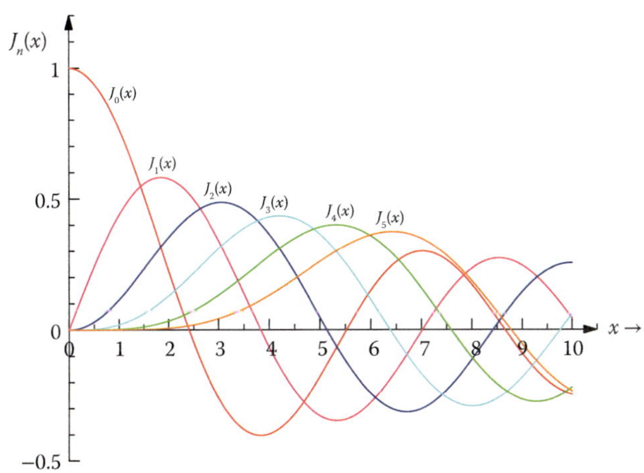

FIGURE 2.6
Bessel function of the first kind $J_n(x)$ for $n = 0, 1, \ldots 5$.

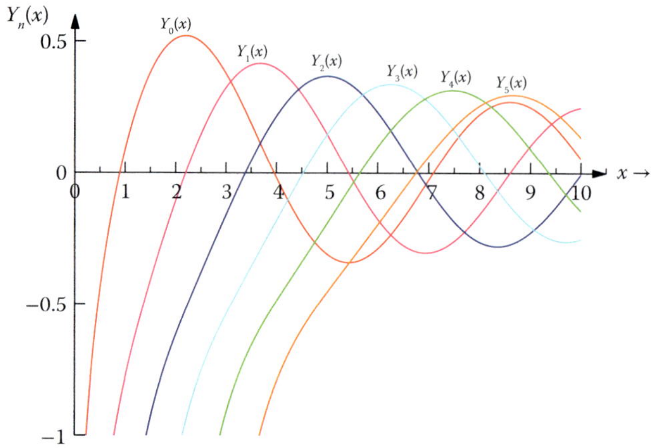

FIGURE 2.7
Bessel function of the second kind $Y_n(x)$ for $n = 0, 1, \ldots 5$.

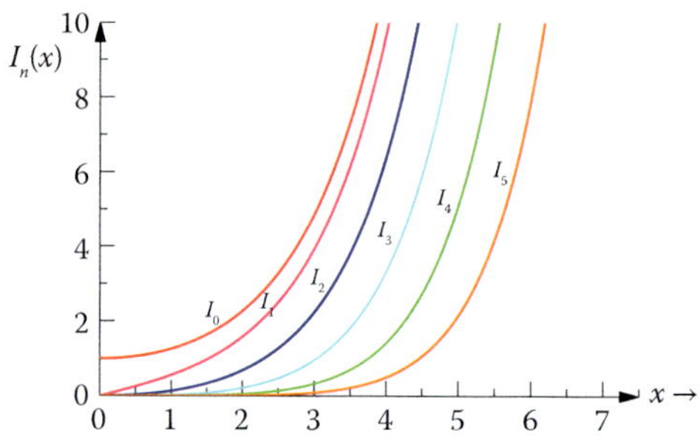

FIGURE 2.8
Modified Bessel function of the first kind $I_n(x)$ for $n = 0, 1, \ldots 4$.

For the Neumann function, $n > 0$,

$$\begin{aligned} Y_n(x) &= \frac{2}{\pi} J_n(x) \ln \frac{\gamma x}{2} - \frac{1}{\pi} \sum_{m=0}^{n-1} \frac{(n-m-1)!(x/2)^{2m-n}}{m!} \\ &\quad - \frac{1}{\pi} \sum_{m=0}^{\infty} \frac{(-1)^m (x/2)^{n+2m}}{m!\Gamma(n+m+1)} [p(m) + p(m+n)] \end{aligned} \tag{2.110}$$

where $\gamma = 1.781$ is Euler's constant and

$$p(m) = \sum_{k=1}^{m} \frac{1}{k}, \qquad p(0) = 0 \tag{2.111}$$

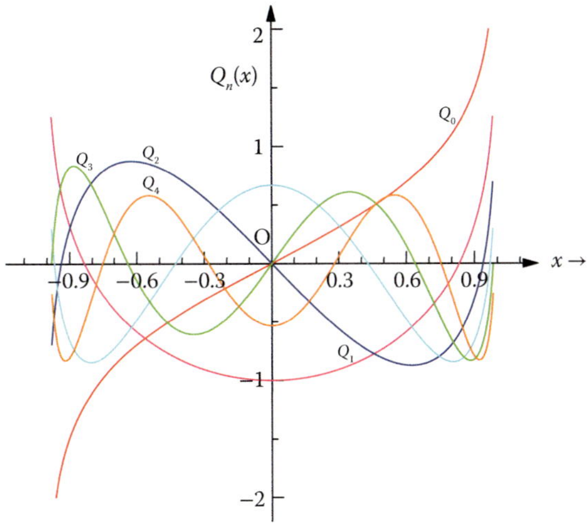

FIGURE 2.9
Modified Bessel function of the first kind $K_n(x)$ for $n = 0, 1, \ldots 4$.

If $n = 0$,

$$Y_0(x) = \frac{2}{\pi} J_0(x) \ln \frac{\gamma x}{2} + \frac{2}{\pi} \sum_{m=0}^{\infty} \frac{(-1)^m (x/2)^{2m}}{(m!)^2} p(m) \tag{2.112}$$

The series expansion of the modified Bessel function of the first kind,

$$I_n(x) = j^{-n} J_n(jx) = \sum_{m=0}^{\infty} \frac{(x/2)^{n+2m}}{m!\,\Gamma(n+m+1)} \tag{2.113}$$

whereas the modified Bessel function of the second kind can be written as

$$K_n(x) = \frac{\pi}{2} \left[J_n(jx) + jY_n(jx) \right] \tag{2.114}$$

If $n > 0$,

$$\begin{aligned} K_n(x) &= \frac{1}{2} \sum_{m=0}^{n-1} \frac{(-1)^n (n-m-1)!(x/2)^{2m-n}}{m!} \\ &\quad + (-1)^n \frac{1}{2} \sum_{m=0}^{\infty} \frac{(x/2)^{2m}}{(m!)^2} p(m) \end{aligned} \tag{2.115}$$

and if $n = 0$,

$$K_0(x) = -I_0(x) \ln \frac{\gamma x}{2} + \sum_{m=0}^{\infty} \frac{(x/2)^{2m}}{(m!)^2} p(m) \tag{2.116}$$

TABLE 2.1
Properties and Identities of Bessel Functions $J_n(x)$

(a)	$J_{-n}(x) = (-1)^n J_n(x)$
(b)	Recurrence formula: $J_n(-x) = (-1)^n J_n(x)$
(c)	$J_{n+1}(x) = \frac{2n}{x} J_n(x) - J_{n-1}(x)$
(d)	$\frac{\mathrm{d}}{\mathrm{d}x} J_n(x) = \frac{1}{2}\left[J_{n-1}(x) - J_{n+1}(x)\right]$
(e)	$\frac{\mathrm{d}}{\mathrm{d}x}\left[x^n J_n(x)\right] = x^n J_{n-1}(x)$
(f)	$\frac{\mathrm{d}}{\mathrm{d}x}\left[x^{-n} J_n(x)\right] = -x^{-n} J_{n-1}(x)$
(g)	$J_n(x) = \frac{1}{\pi}\int_0^{\pi} \cos(n\theta - x\sin\theta)\mathrm{d}\theta, \qquad n \geq 0$
(h)	Fourier–Bessel expansion of $f(x)$: $f(x) = \sum_{k=1}^{\infty} A_k J_n(\lambda_k x)\mathrm{d}x, \quad n \geq 0$; $A_k = \frac{2}{\left[aJ_{n+1}(\lambda_k a)\right]^2}\int_0^a x f(x) J_n(\lambda_k x)\mathrm{d}x, \quad 0 < x < a$; where λ_k's are positive roots in ascending order of magnitude of $J_n(\lambda a) = 0$
(i)	$\int_0^a x J_n(\lambda_i x) J_n(\lambda_j x)\mathrm{d}x = \frac{a^2}{2}\left[J_{n+1}(\lambda_i a)\right]^2 \delta_{ij}$ where λ_i and λ_j are positive roots in ascending order of magnitude of $J_n(\lambda a) = 0$

2.3.2 Diffusion Equation

To be concrete, consider the skin effect on a solid cylindrical conductor. The current density distribution within a good conducting wire $\frac{\sigma}{\omega\epsilon}$ obeys the diffusion equation

$$\nabla^2 \mathbf{J} = \mu\sigma \frac{\partial \mathbf{J}}{\partial t} \tag{2.117}$$

We want to solve this equation for a long conducting wire of radius a. Assuming time-harmonic field with time factor $\mathrm{e}^{j\omega t}$,

$$\nabla^2 \mathbf{J}_s = j\omega\mu\sigma \mathbf{J}_s \tag{2.118}$$

For an infinitely long wire, Equation 2.118 reduces to a one-dimensional problem in cylindrical coordinates:

$$\frac{1}{\rho}\frac{\partial}{\partial \rho}\left(\rho \frac{\partial J_{zs}}{\partial \rho}\right) = j\omega\mu\sigma J_{zs}$$

or

$$\rho^2 J''_{zs} + \rho J'_{zs} - j\omega\mu\sigma J_{zs} = 0 \tag{2.119}$$

Comparing this with Equation 2.107 shows that the solution of Equation 2.119 is the modified Bessel function of zero order. Hence,

$$J_{zs}(\rho) = c_1 I_0(\lambda\rho) + c_2 K_0(\lambda\rho) \tag{2.120}$$

where c_1 and c_2 are constants and

$$\lambda = \sqrt{j\omega\mu\sigma} = \frac{\sqrt{2j}}{\delta} \tag{2.121}$$

and $\delta = \sqrt{\frac{2}{\omega\mu\sigma}}$ is the skin depth. Constant c_2 must vanish if J_z is to be finite at $\rho = 0$. At $\rho = a$,

$$J_{zs}(a) = c_1 I_0(\lambda a) \qquad \rightarrow \qquad c_1 = \frac{J_{zs}(a)}{I_0(\lambda a)}$$

Thus,

$$J_{zs}(\rho) = J_{zs}(a)\frac{I_0(\lambda\rho)}{I_0(\lambda a)} \tag{2.122}$$

If we let $\lambda\rho = \sqrt{2j}\frac{\rho}{\delta} = \sqrt{j}x$, it is convenient to replace

$$I_0(\lambda\rho) = I_0(\sqrt{j}x) = J_0(xe^{j3\pi/4}) = \text{ber}_0(x) + \text{j bei}_0(x) \tag{2.123}$$

where ber_0 and bei_0 are *Kelvin functions* of zero order defined by

$$\text{ber}_0(x) = \sum_{m=0}^{\infty} \frac{\cos(m\pi/2)(x/2)^{2m}}{(m!)^2} \tag{2.124}$$

$$\text{bei}_0(x) = \sum_{m=0}^{\infty} \frac{\sin(m\pi/2)(x/2)^{2m}}{(m!)^2} \tag{2.125}$$

Using ber and bei functions, Equation 2.122 may be written as

$$J_{zs}(\rho) = J_{zs}(a)\frac{\text{ber}_0(x) + \text{j bei}_0(x)}{\text{ber}_0(y) + \text{j bei}_0(y)} \tag{2.126}$$

where $x = \sqrt{2}\rho/\delta$, and $y = \sqrt{2}a/\delta$.

2.3.3 Wave Equation

Assuming time factor $e^{j\omega t}$, we now solve the wave equation

$$\frac{1}{\rho}\frac{\partial}{\partial\rho}\left(\rho\frac{\partial U}{\partial\rho}\right) + \frac{1}{\rho^2}\frac{\partial^2 U}{\partial\varphi^2} + \frac{\partial^2 U}{\partial z^2} + k^2 U = 0 \tag{2.127}$$

Let

$$U(\rho, \varphi, z) = R(\rho)F(\varphi)Z(z) \tag{2.128}$$

Substituting Equation 2.128 into Equation 2.127 and dividing by RFZ/ρ^2 yields

$$\frac{\rho}{R}\frac{\partial}{\partial\rho}\left(\rho\frac{\partial R}{\partial\rho}\right) + \frac{\rho^2}{Z}\frac{\partial^2 Z}{\partial z^2} + k^2\rho^2 = -\frac{1}{F}\frac{\partial^2 F}{\partial\varphi^2} = n^2$$

where $n = 0, 1, 2, \ldots$ and n^2 is the separation constant. Thus,

$$F'' + n^2 F = 0 \tag{2.129}$$

$$\frac{\rho}{R}\frac{\partial}{\partial\rho}\left(\rho\frac{\partial R}{\partial\rho}\right) + \frac{\rho^2}{Z}\frac{\partial^2 Z}{\partial z^2} + k^2\rho^2 = n^2 \tag{2.130}$$

Dividing both sides of Equation 2.130 by ρ^2 leads to

$$\frac{1}{\rho R}\frac{\partial}{\partial\rho}\left(\rho\frac{\partial R}{\partial\rho}\right) + \left(k^2 - \frac{n^2}{\rho^2}\right) = -\frac{1}{Z}\frac{\partial^2 Z}{\partial z^2} = \mu^2$$

where μ^2 is another separation constant. Hence,

$$-\frac{1}{Z}\frac{\partial^2 Z}{\partial z^2} = \mu^2 \tag{2.131}$$

$$\frac{1}{\rho R}\frac{\partial}{\partial\rho}\left(\rho\frac{\partial R}{\partial\rho}\right) + \left(k^2 - \mu^2 - \frac{n^2}{\rho^2}\right) = 0 \tag{2.132}$$

If we let

$$\lambda^2 = k^2 - \mu^2$$

the three separated Equations 2.129, 2.131, and 2.132 become

$$F'' + n^2 F = 0 \tag{2.133}$$

$$Z'' + \mu^2 Z = 0 \tag{2.134}$$

$$\rho^2 R'' + \rho R' + \left[\lambda^2\rho - n^2\right] R = 0 \tag{2.135}$$

The solution to Equation 2.133 is given by

$$F(\varphi) = c_1 e^{jn\varphi} + c_2 e^{-jn\varphi} \tag{2.136a}$$

or

$$F(\varphi) = c_3 \cos n\varphi + c_4 \sin n\varphi \tag{2.136b}$$

Similarly, Equation 2.134 has the solution

$$Z(z) = c_5 e^{j\mu z} + c_6 e^{-j\mu z} \tag{2.137a}$$

or

$$Z(z) = c_7 \cos \mu z + c_8 \sin \mu z \tag{2.137b}$$

To solve Equation 2.135, we let $x = \lambda\rho$ and replace R by y; $R' = \lambda y'$ and $R'' = \lambda^2 y''$ and Equation 2.135 becomes

$$x^2 y'' + xy\prime + (x^2 - n^2)y = 0 \tag{2.138}$$

This is Bessel's differential equation with the general solution

$$y(x) = b_1 J_n(x) + b_2 Y_n(x) \tag{2.139}$$

Any of the Bessel functions or related functions can be a solution of Equation 2.138. We already mentioned Bessel modified functions $I_n(x)$ and $K_n(x)$. Other functions closely related to Bessel functions are *Hankel functions* of the first and second kinds defined, respectively, as

$$H_n^{(1)} = J_n(x) + jY_n(x) \tag{2.140a}$$

$$H_n^{(2)} = J_n(x) - jY_n(x) \tag{2.140b}$$

Hankel functions are analogous to functions $\exp(\pm jx)$ just as J_n and Y_n are analogous to cosine and sine functions. It is possible to obtain the asymptotic expressions for these functions as

$$J_n(x) \xrightarrow{x\to\infty} \sqrt{\frac{2}{\pi x}}\cos(x - n\pi/2 - \pi/2) \tag{2.141a}$$

$$Y_n(x) \xrightarrow{x\to\infty} \sqrt{\frac{2}{\pi x}}\sin(x - n\pi/2 - \pi/2) \tag{2.141b}$$

$$H_n^{(1)}(x) \xrightarrow{x\to\infty} \sqrt{\frac{2}{\pi x}}\exp j(x - n\pi/2 - \pi/2) \tag{2.141c}$$

$$H_n^{(2)}(x) \xrightarrow{x\to\infty} \sqrt{\frac{2}{\pi x}}\exp -j(x - n\pi/2 - \pi/2) \tag{2.141d}$$

$$I_n(x) \xrightarrow{x\to\infty} \sqrt{\frac{1}{2\pi x}}\exp(x) \tag{2.141e}$$

$$K_n(x) \xrightarrow{x\to\infty} \sqrt{\frac{1}{2\pi x}}\exp(-x) \tag{2.141f}$$

With the time factor $e^{j\omega t}$, $H_n^{(1)}(x)$ and $H_n^{(2)}(x)$ represent inward and outward traveling waves, respectively, while $J_n(x)$ or $Y_n(x)$ represents a standing wave. With the time factor $\mathrm{e}^{-j\omega t}$, the roles of $H_n^{(1)}(x)$ and $H_n^{(2)}(x)$ are reversed. For further treatment of Bessel and related functions, refer to the work of Watson [3] and Bell [4]. If we choose $R(\rho) = J_n(x) = J_n(\lambda\rho)$ with Equations 2.136 and 2.137 and apply the superposition theorem, the solution to Equation 2.127 is

$$U(\rho, \varphi, z) = \sum_n \sum_\mu A_{n\mu} J_n(\lambda\rho)\exp(\pm jn\varphi \pm j\mu z) \tag{2.142}$$

Introducing the time factor $e^{j\omega t}$, we finally get

$$\Phi(\rho, \varphi, z, t) = \sum_{m}\sum_{n}\sum_{\mu} A_{n\mu} J_n(\lambda\rho) \exp\left(\pm jn\varphi \pm j\mu z \pm j\omega t\right) \tag{2.143}$$

where $\omega = kc$.

2.4 Spherical Coordinates

Spherical coordinates (r, θ, φ) may be defined as in Figure 2.10. The ranges of the coordinates are: $0 \le r \le \infty$, $\quad 0 \le \theta \le \pi$, $\quad 0 \le \varphi \le 2\pi$. As usual, we will first solve Laplace's equation in two dimensions. Later we solve heat equation and wave equation.

2.4.1 Laplace's Equation

Consider the problem of finding the potential distribution due to an uncharged conducting sphere of radius a located in an external uniform electric field as in Figure 2.11.

The external electric field can be described as

$$\mathbf{E} = E_0\mathbf{a}_z \tag{2.144}$$

while the corresponding electric potential can be described as

$$V = -\int \mathbf{E} \cdot \mathbf{dl} = -E_0 z$$

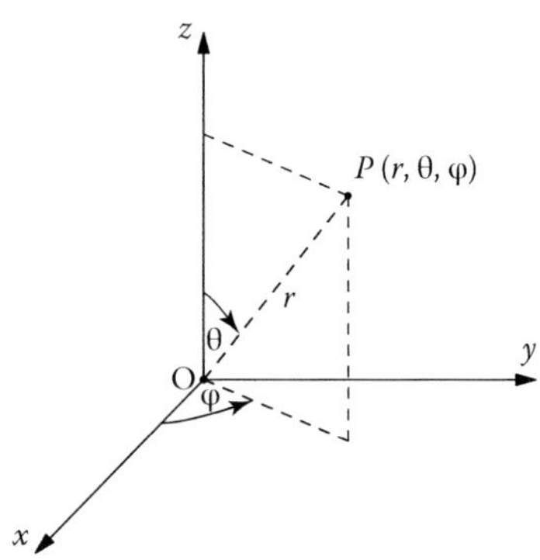

FIGURE 2.10
Spherical coordinates.

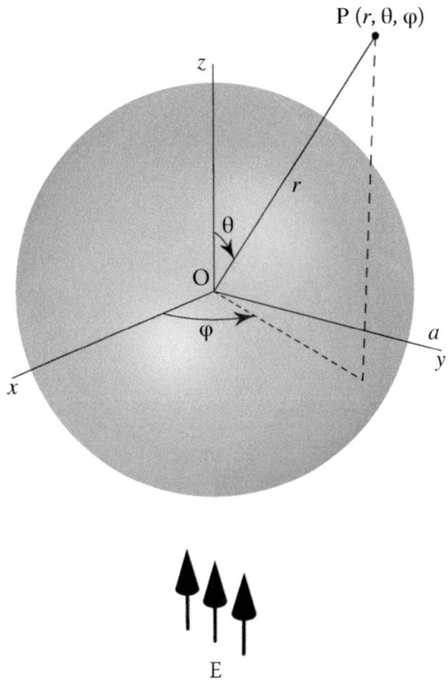

FIGURE 2.11
An uncharged conducting sphere in a uniform external electric field.

or

$$V = -E_0 r \cos\theta \tag{2.145}$$

where $V(\theta = \pi/2) = 0$ has been assumed. From Equation 2.145, it is evident that V is independent of φ, and hence our problem is solving Laplace's equation in two dimensions, namely,

$$\nabla^2 V = \frac{1}{r^2}\frac{\partial}{\partial r}\left(r^2\frac{\partial V}{\partial r}\right) + \frac{1}{r^2\sin\theta}\frac{\partial}{\partial\theta}\left(\sin\theta\frac{\partial V}{\partial\theta}\right) = 0 \tag{2.146}$$

subject to the conditions

$$V(r,\theta) = -E_0 r\cos\theta \qquad \text{as} \qquad r\to\infty \tag{2.147a}$$

$$V(a,\theta) = 0 \tag{2.147b}$$

Using the SoV, we look for a solution of the form

$$V(r,\theta) = R(r)H(\theta) \tag{2.148}$$

so that Equation 2.146 becomes

$$\frac{1}{R}\frac{\mathrm{d}}{\mathrm{d}r}\left(r^2R'\right) = -\frac{1}{H\sin\theta}\frac{\mathrm{d}}{\mathrm{d}\theta}\left(\sin\theta H'\right) = \lambda \tag{2.149}$$

where λ is the separation constant. We shall replace λ with $n(n+1)$ for reasons to be evident later. The separated equations then can be written as

$$r^2R'' + 2rR' - n(n+1)R = 0 \tag{2.150}$$

and

$$\frac{\mathrm{d}}{\mathrm{d}\theta}(\sin\theta H') + \lambda\sin\theta H = 0 \tag{2.151}$$

Equation 2.150 is the *Cauchy–Euler equation.* It can be solved by making the substitution $r = \mathrm{e}^u$, which provides

$$\begin{aligned}\frac{\mathrm{d}R}{\mathrm{d}r} &= \frac{\mathrm{d}R}{\mathrm{d}u}\frac{\mathrm{d}u}{\mathrm{d}r} = \mathrm{e}^{-u}\frac{\mathrm{d}R}{\mathrm{d}u}\\ \frac{\mathrm{d}^2R}{\mathrm{d}\rho^2} &= \mathrm{e}^{-u}\frac{\mathrm{d}}{\mathrm{d}u}\left(\mathrm{e}^{-u}\frac{\mathrm{d}R}{\mathrm{d}u}\right)\\ &= \mathrm{e}^{-2u}\left(\frac{\mathrm{d}^2R}{\mathrm{d}u^2} - \frac{\mathrm{d}R}{\mathrm{d}u}\right)\end{aligned}$$

With these substitutions Equation 2.150 becomes

$$\frac{\mathrm{d}^2R}{\mathrm{d}u^2} + \frac{\mathrm{d}R}{\mathrm{d}u} - n(n+1)R = 0$$

whose solution for a specific value of n

$$R_n(u) = A_n\mathrm{e}^{nu} + \frac{B_n}{\mathrm{e}^{(n+1)u}}$$

This leads to the solution

$$R_n(r) = A_nr^n + \frac{B_n}{r^{n+1}}, \qquad n = 0,\ 1,\ 2,\ \ldots \tag{2.152}$$

To solve Equation 2.151 , we may replace H by y and $\cos\theta$ by x so that

$$\begin{aligned}\frac{\mathrm{d}}{\mathrm{d}\theta} = \frac{\mathrm{d}x}{\mathrm{d}\theta}\frac{\mathrm{d}}{\mathrm{d}x} &= -\sin\theta\frac{\mathrm{d}}{\mathrm{d}x}\\ \frac{\mathrm{d}}{\mathrm{d}\theta}\left(\sin\theta\frac{\mathrm{d}H}{\mathrm{d}\theta}\right) &= -\sin\theta\frac{\mathrm{d}}{\mathrm{d}x}\left(\sin\theta\frac{\mathrm{d}x}{\mathrm{d}\theta}\frac{\mathrm{d}H}{\mathrm{d}x}\right)\\ &= \sin\theta\frac{\mathrm{d}}{\mathrm{d}x}\left(\sin^2\theta\frac{\mathrm{d}y}{\mathrm{d}x}\right)\\ &= \sqrt{1-x^2}\frac{\mathrm{d}}{\mathrm{d}x}\left((1-x^2)\frac{\mathrm{d}y}{\mathrm{d}x}\right)\end{aligned} \tag{2.153}$$

Making these substitutions in Equation 2.151 yields

$$\frac{\mathrm{d}}{\mathrm{d}x}\left[(1-x^2)\frac{\mathrm{d}y}{\mathrm{d}x}\right] + n(n+1)y = 0$$

or

$$(1 - x^2)y'' - 2xy' + n(n+1)y = 0 \tag{2.154}$$

which is the *Legendre differential equation*. Its solution is obtained by the method of Frobenius [2] as

$$y(x) = c_n P_n(x) + d_n Q_n(x) \tag{2.155}$$

where $P_n(x)$ and $Q_n(x)$ are Legendre functions of the first and second kind, respectively. The Legendre function of the first kind is given by

$$P_n(x) = \sum_{k=0}^{N} \frac{(-1)^k (2n-2k)!}{x}^{n-2k} \quad 2^n k!(n-k)!(n-2k)! \tag{2.156}$$

where $N = n/2$ if n is even and $N = (n-1)/2$ if n is odd. For example,

$$\begin{aligned}
P_0(x) &= 1 \\
P_1(x) &= x = \cos\theta \\
P_2(x) &= \frac{1}{2}(3x^2 - 1) = \frac{1}{4}(3\cos 2\theta + 1) \\
P_3(x) &= \frac{1}{2}(5x^3 - 3x) = \frac{1}{8}(5\cos 3\theta + 3\cos\theta) \\
P_4(x) &= \frac{1}{8}(35x^4 - 30x^2 + 3) = \frac{1}{64}(35\cos 4\theta + 20\cos 2\theta + 9) \\
P_5(x) &= \frac{1}{8}(63x^5 - 70x^3 + 15x) = \frac{1}{128}(63\cos 5\theta + 35\cos 3\theta + 30\cos\theta)
\end{aligned} \tag{2.157}$$

Some useful identities and properties of Legendre functions are listed in Table 2.2. The Legendre functions of the second kind are given by

$$Q_n(x) = P_n(x)\left[\frac{1}{2}\ln\left(\frac{1+x}{1-x}\right) - p(n)\right] + \sum_{k=1}^{n} \frac{(-1)^k (n+k)!}{(k!)^2 (n-k)!} p(k) \left[\frac{1-x}{2}\right]^k \tag{2.158}$$

where $p(k)$ is defined in Equation 2.111. Typical graphs of $P_n(x)$ and $Q_n(x)$ are shown in Figures 2.12 and 2.13. The functions Q_n are not as useful as P_n since they are singular at $x = \pm 1$ or ($\theta = 0,\ \pi$) due to the logarithmic term in Equation 2.158. We use Q_n only when $x \neq \pm 1$ or ($\theta \neq 0,\ \pi$), for example, in problems having conical boundaries that exclude the axis from the solution region.

For the problem at hand, ($\theta = 0,\ \pi$) is included so that the solution to Equation 2.151 is

$$H_n(\theta) = P_n(\cos\theta) \tag{2.159}$$

Substituting Equations 2.152 and 2.159 into Equation 2.148 gives

$$V(r, \theta) = \left[A_n r^n + \frac{B_n}{r^{n+1}}\right] P_n(\cos\theta) \tag{2.160}$$

TABLE 2.2
Properties and Identities of Legendre Functions $J_n(x)$

(a)	For $n \geq 1$, $P_n(1) = 1$, $P_n(-1) = (-1)^n$ $P_{2n+1}(0) = 0$, $P_{2n}(0) = (-1)^n \frac{(2n)!}{2^{2n}(n!)^2}$
(b)	$P_n(-x) = (-1)^n P_n(x)$
(c)	$P_n(x) = \frac{1}{2^n n!} \frac{d^n}{dx^n}(x^2 - 1)^n$, $n \geq 0$ Rodriguez formula
(d)	$(n+1)P_{n+1}(x) = (2n+1)xP_n(x) - nP_{n-1}(x)$, $n \geq 1$ Recurrence relation
(e)	$P_n(x) = xP_{n-1}(x) + nP_{n-1}(x)$, $n \geq 1$
(f)	$P_n(x) = xP_{n-1}(x) + \frac{x^2-1}{n} P_{n-1}(x)$, $n \geq 1$
(g)	$P_{n+1}(x) - P_{n-1}(x) = (2n+1)P_n(x)$, $n \geq 1$ or $\int P_n(x)dx = \frac{P_{n+1}(x) - P_{n-1}(x)}{2n+1}$
(h)	Fourier–Legendre series expansion of $f(x)$: $f(x) = \sum_{n=0}^{\infty} A_n P_n(x) dx$, $-1 \leq x \leq 1$ $A_n = \frac{2n+1}{2} \int_{-1}^{1} f(x) P_n(x) dx$, $n \geq 0$ If $f(x)$ is odd, $A_n = (2n+1) \int_0^1 f(x) P_n(x) dx$, $n = 0, 2, 4, \ldots$ If $f(x)$ is even, $A_n = (2n+1) \int_0^1 f(x) P_n(x) dx$, $n = 1, 3, 5, \ldots$
(i)	$\int_{-1}^{1} P_n(x) P_m(x) dx = \begin{cases} 0, & n \neq m \\ \frac{2}{2n+1}, & n = m \end{cases}$

To determine A_n and B_n, we apply the boundary conditions in Equation 2.147. Since as $r \to \infty$, $V = -E_0 r \cos\theta$, it follows that $n = 1$ and $A_1 = -E_0$; that is,

$$V(r, \theta) = \left(-E_0 r + \frac{B_1}{r^2}\right) \cos\theta \tag{2.161}$$

Also since $V = 0$ when $r = a$, $B_1 = E_0 a^3$. Hence, the complete solution is

$$V(r, \theta) = -E_0 \left(r - \frac{a^3}{r^2}\right) \cos\theta \tag{2.162}$$

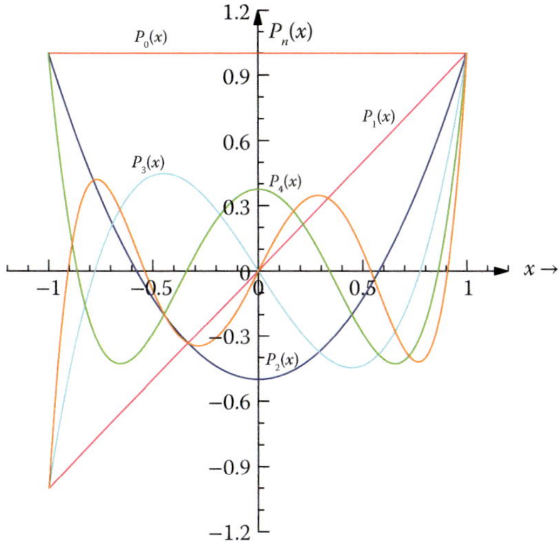

FIGURE 2.12
Typical graphs of $P_n(x)$ for $n = 0, 1, \ldots 4$.

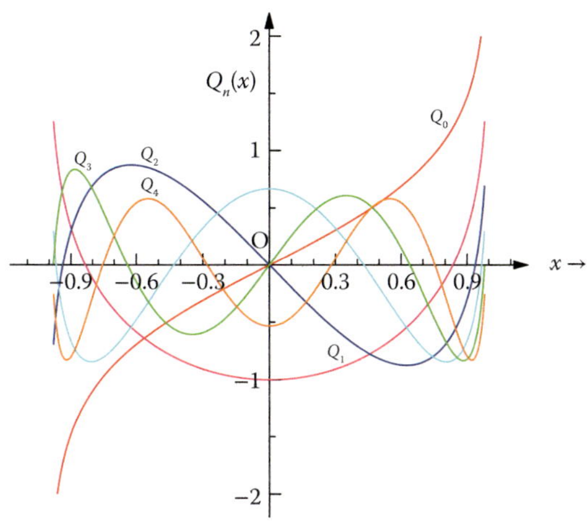

FIGURE 2.13
Typical graphs of $Q_n(x)$ for $n = 0, 1, \ldots 4$.

The electric field intensity is given by

$$\begin{aligned}\mathbf{E} = -\nabla V &= \frac{\partial V}{\partial r}\mathbf{a_r} - \frac{1}{r}\frac{\partial V}{\partial \theta}\mathbf{a}_\theta \\ &= E_0\left(1 + \frac{2a^3}{r^3}\right)\cos\theta\mathbf{a_r} + E_0\left(1 - \frac{a^3}{r^3}\right)\sin\theta\mathbf{a}_\theta \end{aligned} \tag{2.163}$$

2.4.2 Heat Equation

Consider a solid sphere of radius a initially at a temperature of $\sqrt{r}F(r,\theta)$ whose surface is maintained at zero temperature. The transient heat equation in two dimensions is given by

$$\frac{\partial^2 U}{\partial r^2} + \frac{2}{r}\frac{\partial U}{\partial r} + \frac{1}{r\sin\theta}\frac{\partial}{\partial\theta}\left(\sin\theta\frac{\partial U}{\partial\theta}\right) = \frac{1}{\alpha}\frac{\partial U}{\partial t} \tag{2.164}$$

where α is a constant. We solve Equation 2.164 subject to

$$U(a,t) = 0 \tag{2.165a}$$

$$U(r,\theta,0) = \sqrt{r}F(r,\theta) \tag{2.165b}$$

Introducing the transformation $x = \cos\theta$ results in

$$\frac{\partial^2 U}{\partial r^2} + \frac{2}{r}\frac{\partial U}{\partial r} + \frac{1}{r}\frac{\partial}{\partial x}\left((1-x^2)\frac{\partial U}{\partial x}\right) = \frac{1}{\alpha}\frac{\partial U}{\partial t} \tag{2.166}$$

We may also introduce the transformation $V = \sqrt{r}U$. This leads to

$$\frac{\partial^2 V}{\partial r^2} + \frac{2}{r}\frac{\partial V}{\partial r} - \frac{1}{4}\frac{V}{r^2} + \frac{1}{r^2}\frac{\partial}{\partial x}\left((1-x^2)\frac{\partial V}{\partial x}\right) = \frac{1}{\alpha}\frac{\partial U}{\partial t} \tag{2.167}$$

As usual, we assume a product solution

$$V(r,x,\varphi,t) = R(r)M(x)T(t) \tag{2.168}$$

The separated equations are

$$T' + \alpha\lambda^2 T = 0 \tag{2.169}$$

with the general solution

$$T(t) = c_1 \mathrm{e}^{-\alpha\lambda^2 t} \tag{2.170}$$

and

$$R'' + \frac{R'}{r} + \left[\lambda^2 - (n+1/2)^2\frac{1}{r^2}\right]R = 0 \tag{2.171}$$

This is Bessel's equation of order $n + 1/2$ with a general solution

$$R(r) = c_2 J_{n+1/2}(\lambda r) + c_3 Y_{n+1/2}(\lambda r) \tag{2.172}$$

More will be said about these Bessel's functions of fractional order in the next subsection. Since $Y_{n+1/2}(\lambda r)$ is singular at the origin, $c_3 = 0$. For $M(x)$, the separated equation is

$$\frac{\mathrm{d}}{\mathrm{d}x}\left[(1-x^2)M'\right] + n(n+1)M = 0$$

or

$$(1 - x^2)M'' - 2xM' + n(n+1)M = 0 \tag{2.173}$$

which is the Legendre differential equation with a general solution

$$M(x) = c_4 P_n(x) + c_5 Q_n(x) \tag{2.174}$$

Since Q_n is singular at $x = 1$, $c_5 = 0$. Combining Equations 2.170, 2.172, and 2.174, we obtain the complete solution as

$$V(x,r,t) = \sum_{n=0}^{\infty}\sum_{p=1}^{\infty} c_{np} \exp\left(-\alpha\lambda_{np}^2 t\right) J_{n+1/2}(\lambda_{np} a) P_n(x) \tag{2.175}$$

where $J_{n+1/2}(\lambda_{np}a)$. The coefficients λ_{np} is determined from the initial condition. We obtain the final result in terms of U as

$$U(r,\theta,t) = \sum_{n=1}^{\infty}\sum_{p=1}^{\infty} A_{np} \exp\left(-\alpha\lambda_{np}^2 t\right) r^{-1/2} J_{n+1/2}(\lambda_{np} r) P_n(\theta) \tag{2.176}$$

where

$$A_{np} = \frac{1}{N(n)B(\lambda_{np})}\int_{r=0}^{a}\int_{x=-1}^{1} r^{3/2} J_{n+1/2}(\lambda_{np} r) P_n(x) F(r,x)\,\mathrm{d}x\,\mathrm{d}r \tag{2.177a}$$

$$N(n) = \frac{2}{2n+1} \tag{2.177b}$$

$$B(\lambda_{np}) = \frac{a^2}{2}\left[J'_{n+1/2}(\lambda_{np} a)\right]^2 \tag{2.177c}$$

2.4.3 Wave Equation

In spherical coordinates, the wave equation is

$$\nabla^2 U + k^2 U = \frac{1}{r^2}\frac{\partial}{\partial r}\left(r^2 \frac{\partial U}{\partial r}\right) + \frac{1}{r^2 \sin\theta}\frac{\partial}{\partial \theta}\left(\sin\theta \frac{\partial U}{\partial \theta}\right) + \frac{1}{r^2 \sin^2\theta}\frac{\partial^2 U}{\partial \varphi^2} + k^2 U = 0 \tag{2.178}$$

We substitute

$$U(r,\theta,\varphi) = R(r)H(\theta)F(\varphi) \tag{2.179}$$

into Equation 2.178. Multiplying the result by $r^2 \sin^2\theta / RHF$ gives

$$\frac{\sin^2\theta}{R}\frac{\mathrm{d}}{\mathrm{d}r}\left(r^2 \frac{\mathrm{d}R}{\mathrm{d}r}\right) + \frac{\sin\theta}{\mathrm{d}\theta}\frac{\mathrm{d}}{\mathrm{d}\theta}\left(\sin\theta \frac{\mathrm{d}H}{\mathrm{d}\theta}\right) + k^2 r^2 \sin^2\theta = -\frac{1}{F}\frac{\mathrm{d}^2 F}{\mathrm{d}\varphi^2} \tag{2.180}$$

Since the left-hand side is independent of φ, we let

$$-\frac{1}{F}\frac{\mathrm{d}^2 F}{\mathrm{d}\varphi^2} = m^2, \qquad m = 0, 1, 2, \ldots \tag{2.181}$$

where m, the first separation constant, is chosen to be a nonnegative integer such that U is periodic in φ. This requirement is necessary for physical reasons that will be evident later. Thus, Equation 2.180 reduces to

$$\frac{1}{R}\frac{\mathrm{d}}{\mathrm{d}r}\left(r^2\frac{\mathrm{d}R}{\mathrm{d}r}\right) + k^2r^2 = -\frac{1}{H\sin\theta}\frac{\mathrm{d}}{\mathrm{d}\theta}\left(\sin\theta\frac{\mathrm{d}H}{\mathrm{d}\theta}\right) + \frac{m^2}{\sin^2\theta} = \lambda \tag{2.182}$$

where λ is the second separation constant. As in Equations 2.150 through 2.152, $\lambda = n(n+1)$ so that the separated equations are now

$$F'' + m^2F = 0 \tag{2.183}$$

$$R'' + \frac{2}{R}R' + \left[k^2 - \frac{n(n+1)}{r^2}\right]R = 0 \tag{2.184}$$

and

$$\frac{1}{\sin\theta}\frac{\mathrm{d}}{\mathrm{d}\theta}(\sin\theta H') + \left[n(n+1) - \frac{m^2}{\sin^2\theta}\right]H = 0 \tag{2.185}$$

As usual, the solution to Equation 2.183 is

$$F(\varphi) = c_1\mathrm{e}^{jm\varphi} + c_2\mathrm{e}^{-jm\varphi} \tag{2.186a}$$

or

$$F(\varphi) = c_3\sin m\varphi + c_4\cos\varphi \tag{2.186b}$$

If we let $R(r) = r^{-1/2}G(r)$, Equation 2.184 becomes

$$G'' + \frac{1}{r}G' + \left[k^2 - \frac{(n+1/2)^2}{r^2}\right]G = 0 \tag{2.187}$$

which has the solution

$$G = Ar^{1/2}z_n(kr) + BZ_{n+1/2}(kr) \tag{2.188}$$

Functions $z_{n+1/2}(x)$ are spherical Bessel functions and are related to ordinary Bessel functions according to

$$z_n(x) = \sqrt{\frac{\pi}{2x}}Z_{n+1/2}(x) \tag{2.189}$$

In Equation 2.189, $Z_{n+1/2}(x)$ may be any of the ordinary Bessel function of half-integer order $J_{n+1/2}(x)$, $Y_{n+1/2}(x)$, $I_{n+1/2}(x)$, $K_{n+1/2}(x)$, $H^{(1)}_{n+1/2}(x)$, and $H^{(2)}_{n+1/2}(x)$, while $z_n(x)$ may be any of the corresponding $j_n(x), y_n(x), i_n(x)$,

$k_n(x)$, $h_n^{(1)}(x)$, and $h_n^{(2)}(x)$. Bessel functions of fractional order are, in general, given by

$$J_\nu(x) = \sum_{k=0}^{\infty} \frac{(-1)^k x^{2k+\nu}}{2^{2k+\nu} k! \Gamma(\nu + k + 1)} \tag{2.190}$$

$$Y_\nu(x) = \frac{J_\nu(x)\cos(\nu\pi) - J_\nu(x)}{\sin(\nu\pi)} \tag{2.191}$$

$$I_\nu(x) = (-j)^\nu J_\nu(jx) \tag{2.192}$$

$$K_v(x) = \frac{\pi}{2}\left[\frac{I_{-\nu}(x) - I_\nu(x)}{\sin \nu\pi}\right] \tag{2.193}$$

where $J_{-\nu}$ and $I_{-\nu}$ are, respectively, obtained from Equations 2.190 and 2.192 by replacing ν and $-\nu$. Although ν in Equations 2.190 through 2.193 can assume any fractional value, in our specific problem, $\nu = n+1/2$. Since Gamma function of half-integer order is needed in Equation 2.190, it is necessary to add that

$$\Gamma(n + 1/2) = \begin{cases} \frac{(2n)!}{2^{2n}\, n!}\sqrt{\pi} & n \geq 0 \\ \\ \frac{(-1)^n\, 2^{2n}\, n!}{(2n)!}\sqrt{\pi} & n < 0 \end{cases} \tag{2.194}$$

Thus, the lower-order spherical Bessel functions are as follows:

$$\begin{aligned}
j_0(x) &= \frac{\sin x}{x} & y_0(x) &= -\frac{\cos x}{x} \\
h_0^{(1)}(x) &= \frac{e^{jx}}{jx} & h_0^{(2)}(x) &= -\frac{e^{-jx}}{-jx} \\
i_0(x) &= \frac{\sinh x}{x} & k_0(x) &= \frac{e^{x}}{x} \\
j_1(x) &= \frac{\sin x}{x^2} - \frac{\cos x}{x} & y_1(x) &= -\frac{\cos x}{x^2} - \frac{\sin x}{x} \\
h_1^{(1)}(x) &= -\frac{x+j}{x^2}e^{jx} & h_1^{(2)}(x) &= -\frac{x-j}{x^2}e^{-jx}
\end{aligned} \tag{2.195}$$

Other $z_n(x)$ can be obtained from the series expansion in Equations 2.190 through 2.193 or the recurrence relations and properties of $z_n(x)$ presented in Table 2.3.

By replacing H in Equation 2.185 with y, $\cos\theta$ by x and making other substitution as we did in Equation 2.153, we obtain

$$(1 - x^2)y'' - 2xy' + \left[n(n+1) - \frac{m^2}{1 - x^2}\right] y = 0 \tag{2.196}$$

which is Legendre's associated differential equation. Its general solution is of the form

$$y(x) = a_{mn}P_n^m(x) + d_{mn}Q_n^m(x) \tag{2.197}$$

TABLE 2.3
Properties and Identities of Spherical Bessel Functions $z_n(x)$

(a)	$z_{n+1}(x) = \frac{2n+1}{x} z_n(x) - z_{n-1}(x)$	Recurrence relation
(b)	$\frac{\mathrm{d}}{\mathrm{d}x} z_n(x) = \frac{1}{2n+1}\left[n z_{n-1}(x) - (n+1) z_{n+1}(x)\right]$	
(c)	$\frac{\mathrm{d}}{\mathrm{d}x}\left[x z_n(x)\right] = -n z_n(x) + x z_{n+1}(x)$	
(d)	$\frac{\mathrm{d}}{\mathrm{d}x}\left[x^{n+1} z_n(x)\right] = -x^{n+1} z_{n-1}(x)$	
(e)	$\frac{\mathrm{d}}{\mathrm{d}x}\left[x^{-n} z_n(x)\right] = -x^{-n} z_{n+1}(x)$	
(f)	$\int x^{n+2} z_n(x)\mathrm{d}x = x^{n+2} z_{n+1}(x)$	
(g)	$\int x^{1-n} z_n(x)\mathrm{d}x = -x^{1-n} z_{n-1}(x)$	
(h)	$\int x^2 \left[z_n(x)\right]^2 \mathrm{d}x = \frac{1}{2} x^3 \left[z_n(x) - z_{n-1}(x)\right]$	

where $P_n^m(x)$ and $Q_n^m(x)$ are called *associated Legendre functions* of the first and second kind, respectively. Equation 2.155 is a special case of Equation 2.197 when $m = 0$. $P_n^m(x)$ and $Q_n^m(x)$ can be obtained from ordinary Legendre functions $P_n(x)$ and $Q_n(x)$ using

$$P_n^m(x) = (1-x^2)^{m/2} \frac{\mathrm{d}^m}{\mathrm{d}x^m} P_n(x) \tag{2.198}$$

$$Q_n^m(x) = (1-x^2)^{m/2} \frac{\mathrm{d}^m}{\mathrm{d}x^m} Q_n(x) \tag{2.199}$$

where $-1 < x < 1$. We note that

$$\begin{aligned} P_n^0(x) &= P_n(x) \\ Q_n^0(x) &= Q_n(x) \\ P_n^m(x) &= 0 \qquad m > n \end{aligned}$$

Typical associated Legendre functions are

$$\begin{aligned} P_1^1(x) &= (1-x^2)^{1/2} = \sin\theta \\ P_2^1(x) &= 3x(1-x^2)^{1/2} = 3\cos\theta\sin\theta \\ P_2^2(x) &= 3(1-x^2) = (3\sin^2\theta) \\ P_3^1(x) &= \frac{3}{2}(1-x^2)^{1/2}(5x-1) = \frac{3}{2}\sin\theta(5\cos 3\theta - 1) \end{aligned} \tag{2.200}$$

$$\begin{aligned} Q_1^1(x) &= -(1-x^2)^{1/2}\left[\frac{1}{2}\ln\frac{1+x}{1-x} + \frac{x}{1-x^2}\right] \\ Q_2^1(x) &= -(1-x^2)^{1/2}\left[\frac{3x}{2}\ln\frac{1+x}{1-x} + \frac{3x^2-2}{1-x^2}\right] \\ Q_2^2(x) &= (1-x^2)^{1/2}\left[\frac{3}{2}\ln\frac{1+x}{1-x} + \frac{5x-3x^2}{(1-x^2)^2}\right] \end{aligned} \tag{2.201}$$

Higher-order associated Legendre functions can be obtained using Equations 2.198 and 2.199 along with the properties in Table 2.4. As mentioned earlier, $Q_n^m(x)$ is unbounded at $x \pm 1$, and hence it is only used when $x \pm 1$ is excluded.

TABLE 2.4
Properties and Identities of Associated Legendre Functions $P_n(x)$

(a) $P_n^m(x) = 0, \qquad m > n$

(b) Recurrence relation for fixed m:

$$P_n^m(x) = \frac{(2n-1)xP_{n-1}^m(x) - (n+m-1)P_{n-2}^m(x)}{n-m}$$

(c) Recurrence relation for fixed n:

$$P_n^m(x) = \frac{2(m-1)x}{(1-x^2)^{1/2}} P_n^{m-1}(x) - (n-m+2)(n+m-1)P_n^{m-2}(x)$$

(d) $P_n^m(x) = \frac{(1-x^2)^{m/2}}{2^n} \sum_{k=0}^{\lceil \frac{m-n}{2} \rceil} \frac{(-1)^k (2n-2k)! x^{n-2k-m}}{k!(n-k)!(n-2k-m)!}$

(e) $\frac{\mathrm{d}}{\mathrm{d}x} P_n^m(x) = \frac{(n+m)P_{n-1}^m(x) - nxP_n^m(x)}{1-x^2}$

(f) $\frac{\mathrm{d}}{\mathrm{d}\theta} P_n^m(x) = \frac{1}{2}\left[(n-m+1)P_n^{m-1}(x) - P_n^{m+1}(x)\right]$

(g) $\frac{\mathrm{d}}{\mathrm{d}x} P_n^m(x) = -\frac{mxP_n^m(x)}{1-x^2} + \frac{(1-x^2)^{m/2}}{2^n} \sum_{k=0}^{\lceil \frac{m-n}{2} \rceil} \frac{(-1)^k (2n-2k)! x^{n-2k-m}}{k!(n-k)!(n-2k-m)!}$

(h) $\frac{\mathrm{d}}{\mathrm{d}\theta} P_n^m x = -(1-x^2)^{1/2} \frac{\mathrm{d}}{\mathrm{d}x} P_n^m(x)$

(i) The series expansion of $f(x)$:

$$f(x) = \sum_{k=1}^{\infty} A_n P_n^m(x)$$

where

$$A_n = \frac{(2n+1)(n-m)!}{2(n+m)!} \int_{-1}^{1} f(x) P_n^m(x)\mathrm{d}x$$

(j) $\frac{\mathrm{d}^m}{\mathrm{d}x^m} P_n(x)\Big|_{x=1} = \frac{(n+m)!}{2^m m!(n-m)!}$

$$\frac{\mathrm{d}^m}{\mathrm{d}x^m} P_n(x)\Big|_{x=-1} = \frac{(-1)^{n+m}(n+m)!}{2^m m!(n-m)!}$$

(k) $P_n^{-m}(x) = (-1)^m \frac{(n-m)!}{(n+m)!} P_n^m(x), \qquad m = 0, 1, \ldots, n$

(l) $\int_{-1}^{1} P_n^m(x) P_k^m(x)\mathrm{d}x = \frac{2}{2n+1} \frac{(n+m)!}{(n-m)!} \delta_{nk}$

where δ_{nk} is the Kronecker delta defined by

$$\delta_{nk} \begin{cases} 0, & n \neq k \\ 1, & n = k \end{cases}$$

Substituting Equations 2.186, 2.188, and 2.197 into Equation 2.179 and applying superposition theorem, we obtain

$$U(r, \theta, \varphi) = \sum_{n=0}^{\infty} \sum_{m=0}^{n} \sum_{\ell=0}^{\infty} A_{mn\ell} z_n(k_{m\ell} r) P_n^m(\cos\theta) \exp(\pm jm\varphi) \qquad (2.202)$$

Note that the products $H(\theta)F(\varphi)$ are known as *spherical harmonics.*

2.5 Conclusion

The method of SoV is developed through the search for exact solutions, which are in closed forms. Its main advantage is the high accuracy of results in comparison with numerical (approximate) methods. It has some weak points. First, the method is applicable to a very restricted set of geometries. The boundary of the solution region must coincide with one of the coordinate surfaces. It cannot handle problems with complicated solution regions. Separation of coordinates may or may not exist. In most practical applications, it is not possible to carry out a SoV. Second, a simple diffraction problem leads to a complicated Fourier series in special functions. For these reasons, SoV is seldom used in practical applications. In this chapter, we have applied SoV for solving PDEs encountered in electromagnetics. SoV can also be used to solve integral equations [5].

References

[1] J. Fourier, *The Analytical Theory of Heat*, Cambridge, UK: University Press, 1878.

[2] T. Myint-U, *Partial Differential Equations of Mathematical Physics*, 2nd ed., New York: North-Holland, 1980, pp. 186–188, 198–204.

[3] G. N. Watson, *Theory of Bessel Functions*, London, UK: Cambridge University Press, 1966.

[4] W. W. Bell, *Special Functions for Scientists and Engineers*, London: D. Van Nostrand, 1968.

[5] T. I. Velytiak and S. A. Yaroshiko, "A method of generalized separation of variables for solving two-dimensional integral equation," *Proceedings of Xth International Seminar/Workshop on Direct and Inverse Problems of Electromagnetic and Acoustic Wave Theory*, 1998, pp. 81–85.

Problems

1. Given the electrostatic potential on the sphere $r = a$ as

$$V(a, \theta) = f(\theta)$$

   (i) Attempt a solution by variable separation technique to solve the Laplace equation $\nabla^2 V = 0$ in spherical coordinates (r, θ).
   (ii) In particular if $f(\theta) = V_0 \cos^2 \theta$, find expressions for the potential $V(r, \theta)$ both inside and outside the sphere.
   (iii) Find the source of surface charge density ρ_s on the sphere that can support this potential function both inside and outside the sphere.

2. Given the PDE and the domain of interest

$$u_{tt} - c^2 u_{xx} = 0, \qquad 0 < x < \ell, \quad t > 0$$

   solve for $u(x, t)$ subject to each of the various boundary and initial conditions listed in Table 2.5.

3. Given the two-dimensional elliptic equation in a rectangular region

$$u_{xx} + u_{yy} - c^2 u = 0, \qquad 0 < x < a, \; 0 < y < b$$

   solve the same subject to the boundary conditions:

$$u(0, y) = u(a, y) = 0, \;\; u(x, 0) = f(x); \;\; u(x, b) = g(x)$$

4. Given the two-dimensional wave equation in a disk

$$u_{tt} - c^2 \left[u_{xx} + u_{yy}\right] = 0, \qquad 0 \leq (x^2 + y^2) < \ell^2, \quad t > 0$$

   solve the same subject to the initial conditions:

$$u(x, y, 0) = f(x, y); \;\; u_t(x, y, 0) = g(x, y)$$

   and the boundary conditions

$$u(x, y, t) = 0; \quad \text{for} \quad x^2 + y^2 = \ell^2; \; t > 0.$$

TABLE 2.5
Initial and Boundary Conditions for Problem 2

(i)	$u(0,t) = u(\ell,t) = u(x,0) = 0;\;\; u_t(x,0) = 1$
(ii)	$u(0,t) = u(\ell,t) = u_t(x,0) = 0;\;\; u(x,0) = x^2(x-\ell)^2$
(iii)	$u(0,t) = u(\ell,t) = u_t(x,0) = 0;\;\; u(x,0) = \sin^2\left(\frac{\pi x}{\ell}\right);$
(iv)	$u(x,0) = 1;\;\; u_t(x,0) = x;\;\; u(0,t) = 0;\;\; u_x(\ell,t) + \eta u(\ell,t) = 0;\quad \eta > 0$

3

Series Expansion Method

Lord, make me an instrument of your peace!
where there is hatred, let me sow love;
where there is injury, pardon;
where there is doubt, faith;
where there is despair, hope;
where there is darkness, light;
where there is sadness, joy.

– St. Francis of Assisi

CONTENTS

Partial differential equations (PDEs) can be solved with the aid of infinite series and more generally with the aid of series of orthogonal functions. In this chapter, we apply the idea of infinite series expansion to those PDEs in which the independent variables are not separable or, if they are separable, the boundary conditions are not satisfied by the particular solutions. After presenting the generalized Fourier series, we apply series expansion to the problems of Poisson's equation in a cube, Poisson's equation in a cylinder, strip transmission line, scattering by a conducting cylinder, and scattering by a dielectric sphere.

3.1 Generalized Fourier Series

Orthogonal functions are of great importance in science and engineering. A system of real functions $\phi_n(x),\ n = 0, 1, 2, \cdots$ is said to be orthogonal with a weighting function $w(x)$ on the interval (a, b) if

$$\int_a^b \phi_m(x)\phi_n(x)w(x)\mathrm{d}x = a_n\delta_{mn} \tag{3.1}$$

where a_n are given constants and δ_{mn} is the Kronecker delta. For example, the system of functions $\cos(nx)$ is orthogonal with weight 1 on the interval $(0, \pi)$ since

$$\int_0^\pi \cos(mx)\cos(nx)\mathrm{d}x = 0, \qquad m \neq n \tag{3.2}$$

Orthogonal functions usually arise in the solution of PDEs governing the behavior of certain physical phenomena. These include Bessel, Legendre, Hermite, Laguerre, and Chebyshev polynomials. They are useful in series expansion of functions to form Fourier–Bessel series, Fourier–Legendre series, and so on. Regular Fourier series deals with periodic functions. Any periodic functions can be expressed in terms of sine and cosine functions. We can generalize that idea for any functions. We can represent almost every analytic function by power series. A function $f(x)$, defined over interval (a, b) can be expressed in terms of any complete, orthogonal set of functions

$$f(x) = \sum_n c_n\phi_n(x) \tag{3.3}$$

where the expansion coefficients are given as

$$c_n = \int_a^b f(x)\phi_n(x)w(x)\mathrm{d}x \tag{3.4}$$

and the (weighted) norm N_n is defined as

$$N_n = \int_a^b \phi_n^2(x)w(x)\mathrm{d}x \tag{3.5}$$

The series in Equation 3.3 is commonly called generalized Fourier series. (It is also called eigenfunction expansion). In this section, we discuss two important series of this type: Fourier–Legendre series and Fourier–Bessel series [1,2]. The Legendre polynomials $P_n(x)$ are orthogonal with respect to the weight function $w(x) = 1$ on $(-1, 1)$. The generalized Fourier–Legendre series associated with the Legendre polynomials is given by

$$f(x) = \sum_n c_nP_n(x), \qquad -1 < x < 1 \tag{3.6}$$

where the Fourier coefficients are given by

$$c_n = \frac{\int_{-1}^{1} f(x)P_n(x)\,\mathrm{d}x}{\int_{-1}^{1} P_n^2(x)\,\mathrm{d}x} = \frac{2n+1}{2}\int_{-1}^{1} f(x)P_n(x)\,\mathrm{d}x \tag{3.7}$$

The generalized Fourier–Bessel series expansion of $f(x)$ that satisfies Dirichlet conditions in the interval $(0, a)$ is given by

$$f(x) = \sum_n c_n J_n(k_n x), \qquad 0 < x < a \tag{3.8}$$

where J_n is the nth order Bessel function of the first kind. The expansion coefficients are

$$c_n = \frac{\int_0^a x f(x) J_n(k_n x)\,\mathrm{d}x}{\int_0^a x J_n^2(k_n x)\,\mathrm{d}x} = \frac{2\int_0^a x f(x) J_n(k_n x)\,\mathrm{d}x}{a^2\left[J_n'(t_n)\right]^2} \tag{3.9}$$

where $t_n = k_n a$ with $J_n(t_n) = 0$.

3.2 Poisson's Equation in a Cube

Consider the problem

$$\nabla^2 V = \frac{\partial^2 V}{\partial x^2} + \frac{\partial^2 V}{\partial y^2} + \frac{\partial^2 V}{\partial z^2} = -f(x, y, z) \tag{3.10}$$

subject to the boundary conditions

$$\begin{aligned} 0 &= V(0, y, z) = V(a, y, z) \\ &= V(x, 0, z) = V(x, b, z) \\ &= V(x, y, 0) = V(x, y, c) \end{aligned} \tag{3.11}$$

where $f(x, y, z)$ the source term, is given. We should note that the independent variables in Equation 3.10 are not separable. However, in Laplace's equation, $f(x, y, z) = 0$ and the variables are separable. Although the problem defined by Equations 3.10 and 3.11 can be solved in several ways, we stress the use of series expansion in this chapter. Let the solution be of the form

$$V(x, y, z) = \sum_{m=1}^{\infty}\sum_{n=1}^{\infty}\sum_{p=1}^{\infty} A_{mnp} \sin\left(\frac{m\pi x}{a}\right) \sin\left(\frac{n\pi y}{b}\right) \sin\left(\frac{p\pi z}{c}\right) \tag{3.12}$$

where the triple sine series is chosen so that the individual terms and the entire series would satisfy the boundary conditions of Equation 3.11. However,

subject to

$$
\begin{aligned}
&V = 0, && \rho = a && \text{(at the wall)} \\
&V_g = V_\ell, && z = b && \text{(at the gas–liquid interface)} \\
&\frac{\partial V_g}{\partial z} = \epsilon \frac{\partial V_\ell}{\partial z}, && z = b && \text{(at the gas–liquid interface)}
\end{aligned}
\tag{3.18}
$$

Applying the series expansion techniques, we let

$$V_\ell = \sum_{n=1}^{\infty} J_n(\lambda_n \rho) F_n(z) \tag{3.19a}$$

$$V_g = \sum_{n=1}^{\infty} J_n(\lambda_n \rho) \left[A_n \sinh(\lambda_n(b + c - z)) + B_n \cosh(\lambda_n(b + c - z))\right] \tag{3.19b}$$

where F_n, A_n, and B_n are to be determined. At $z = b + c$, $V_g = 0$, which implies that $B_n = 0$. Hence, Equation 3.19b becomes

$$V_g = \sum_{n=1}^{\infty} A_n J_0(\lambda_n \rho) \sinh(\lambda_n(b + c - z)) \tag{3.20}$$

Substituting Equation 3.19a into Equation 3.17a yields

$$\sum_{n=1}^{\infty} J_0(\lambda_n \rho) \left[F_n' - \lambda_n^2 F_n\right] = -\frac{\rho_v}{\epsilon}$$

If we let $F_n' - \lambda_n^2 F_n = G_n$, then

$$\sum_{n=1}^{\infty} G_n J_0(\lambda_n \rho) = -\frac{\rho_v}{\epsilon} \tag{3.21}$$

At $\rho = a$, $V_g = V_\ell = 0$, which makes

$$J_0(\lambda_n a) = 0 \tag{3.22}$$

indicating that λ_n are the roots of J_0 divided by a. Multiplying Equation 3.21 by $\rho J_0(\lambda_m \rho)$ and integrating over the interval $0 < \rho < a$ gives

$$\sum_{n=1}^{\infty} G_n \int_0^a J_0(\lambda_n \rho) \rho\, J_0(\lambda_m \rho) \mathrm{d}\rho = -\frac{\rho_v}{\epsilon} \int_0^a J_0(\lambda_m \rho) \mathrm{d}\rho \tag{3.23}$$

The left-hand side is zero except when $m = n$. Also,

$$\int_0^a \rho J_0^2(\lambda_n \rho) \mathrm{d}\rho = \frac{a^2}{2}\left[J_0^2(\lambda_n a) + J_1^2(\lambda_n a)\right] = \frac{a^2}{2} J_1^2(\lambda_n a) \tag{3.24}$$

since $J_0(\lambda_n a) = 0$. Further,

$$\int_0^a \rho J_0(\lambda_m \rho) d\rho = \frac{a}{\lambda_m} J_1(\lambda_m a) \tag{3.25}$$

Substituting Equations 3.24 and 3.25 in Equation 3.23, with n in place of m, we obtain

$$G_n \frac{a^2}{2} J_1^2(\lambda_n a) = -\frac{\rho_v}{\epsilon} \frac{a}{\lambda_n} J_1(\lambda_n a)$$

or

$$G_n = -\frac{2\rho_v}{\epsilon a J_1(\lambda_n a)} \tag{3.26}$$

showing that G is a constant. Thus,

$$F_n' - \lambda_n^2 F_n = G_n \tag{3.27}$$

which is an inhomogeneous ordinary differential equation. Its solution is

$$F_n(z) = C_n \sinh \lambda_n z + D_n \cosh \lambda_n z - \frac{G_n}{\lambda_n^2} \tag{3.28}$$

But

$$F_n(0) = 0 \qquad \to \qquad D_n = \frac{G_n}{\lambda_n^2} \tag{3.29}$$

Thus, Equation 3.19a becomes

$$V_\ell = \sum_{n=1}^{\infty} J_0(\lambda_n \rho) \left[C_n \sinh \lambda_n z + \frac{G_n}{\lambda_n^2} \{\cosh \lambda_n z - 1\} \right] \tag{3.30}$$

Imposing the conditions at $z = b$; that is,

$$V_\ell(\rho, b) = V_g(\rho, b)$$

we obtain

$$A_n \sinh(\lambda_n c) = C_n \sinh \lambda_n z + \frac{G_n}{\lambda_n^2} \{\cosh \lambda_n z - 1\} \tag{3.31}$$

Also, imposing the condition

$$\left.\frac{\partial V_g}{\partial z}\right|_{z=b} = \left.\frac{\partial V_\ell}{\partial z}\right|_{z=b}$$

gives

$$\lambda_n A_n \cosh(\lambda_n c) = -\epsilon_r \lambda_n C_n \cosh \lambda_n b - \frac{\epsilon_r G_n}{\lambda_n^2} \sinh \lambda_n z \tag{3.32}$$

Solving Equations 3.31 and 3.32, we get

$$A_n = \frac{2\rho_v}{R_n K_n}\left[\cosh\left(\lambda_n b\right) - 1\right] \tag{3.33a}$$

$$C_n = \frac{2\rho_v}{R_n \epsilon_r}\left[\cosh\left(\lambda_n b\right)\cosh\left(\lambda_n c\right) + \epsilon_r \sinh\left(\lambda_n b\right)\sinh\left(\lambda_n c\right) - \cosh\lambda_n c\right] \tag{3.33b}$$

where

$$K_n = \sinh\left(\lambda_n b\right)\cosh\left(\lambda_n b\right) + \epsilon_r \cosh\left(\lambda_n b\right)\sinh\left(\lambda_n b\right) \tag{3.34a}$$

$$R_n = \epsilon_0 a \lambda_n^3 J_1(\lambda_n a) \tag{3.34b}$$

Substituting A_n and C_n in Equations 3.20 and 3.30, we obtain the complete solution as

$$V_\ell = \sum_{n=1}^{\infty} \frac{2\rho_v}{R_n \epsilon_r} J_0(\lambda_n \rho)\left[\{\cosh\left(\lambda_n b\right)\cosh\left(\lambda_n c\right) \right. \tag{3.35a}$$

$$\frac{\sinh\left(\lambda_n z\right)}{K_n} \times$$

$$\left. + \epsilon_r \sinh\left(\lambda_n b\right)\sinh\left(\lambda_n c\right) - \cosh\left(\lambda_n c\right)\}\right]$$

$$- \cosh\left(\lambda_n z\right) + 1$$

$$V_g = \sum_{n=1}^{\infty} \frac{2\rho_v}{R_n K_n} J_0(\lambda_n \rho)\left\{\cosh\left(\lambda_n b\right) - 1\right\}\sinh\left(\lambda_n (b + c - z)\right) \tag{3.35b}$$

3.4 Strip Transmission Line

Consider a strip conductor enclosed in a shield box containing homogeneous medium as shown in Figure 3.2a. If TEM mode of propagation is assumed, our problem is reduced to finding V satisfying Laplace's equation $\nabla^2 V = 0$. Due to symmetry, we need to only consider one quarter-section of the line as in Figure 3.2b. This quadrant can be subdivided into regions 1 and 2, where region 1 is under the center conductor and region 2 is not. We now seek solutions V_1 and V_2 for regions 1 and 2, respectively.

If $w = a$, region 1 is similar to a parallel-plate capacitor problem. Thus, we have a one-dimensional problem similar to Equation 2.21 with solution

$$V_1 = a_1 y + a_2 \tag{3.36}$$

Since $V_1(y = 0) = V_0$ and $V_1(y = -b/2) = 0$, $a_2 = V_0$, $a_1 = -2V_0/b$. Hence,

$$V_1(x, y) = V_0 - \frac{2V_0}{b} y \tag{3.37}$$

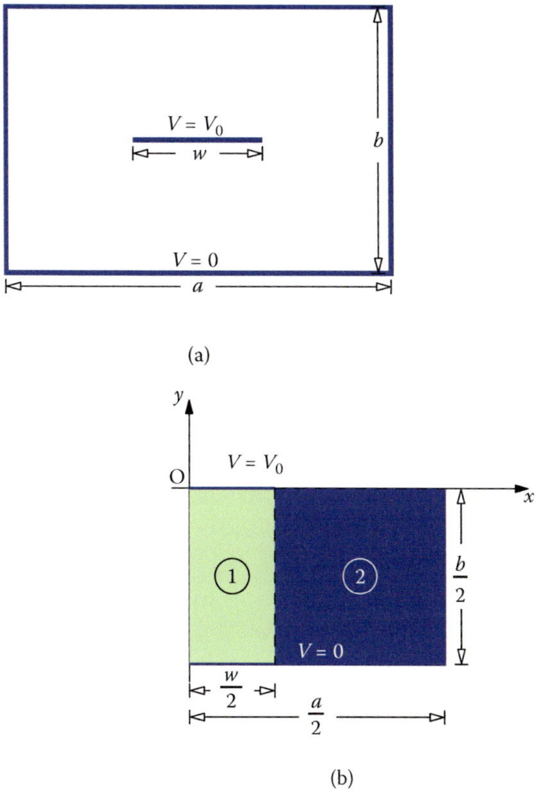

FIGURE 3.2
Strip transmission line. (a) Entire cross-section, (b) A quarter of the cross-section with regions ① and ②.

For region 2, the series expansion solution is of the form

$$V_2(x, y) = \sum_{n=1,3,5}^{\infty} A_n \sin\left(\frac{n\pi y}{b}\right) \sinh\left(\frac{n\pi(a/2 - x)}{b}\right) \tag{3.38}$$

which satisfies Laplace's equation and the boundary conditions along the box. Notice that the even-numbered terms could not be included because they do not satisfy the boundary condition requirements about line $y = 0$; that is, $E_y(y = 0) = -\frac{\partial V_2}{\partial y}\Big|_{y=0} \neq 0$. To determine the expansion coefficients A_n in Equation 3.38, we utilize the fact that V must be continuous at the interface $x = w/2$ between regions 1 and 2; that is,

$$V_1(x = w/2,\, y) = V_2(x = w/2,\, y)$$

or

$$-\frac{2V_0 y}{b} = \sum_{n=odd}^{\infty} A_n \sin\left(\frac{n\pi y}{b}\right) \sinh\left(\frac{n\pi(a - x)}{2b}\right) \tag{3.39}$$

which is the Fourier series expansion. Thus,

$$A_n \sinh\left(\frac{n\pi(a-w)}{2b}\right) = -\frac{2}{b}\int_{-b/2}^{b/2} \frac{2V_0 y}{b}\sin\left(\frac{n\pi y}{b}\right)\mathrm{d}y = \frac{8V_0\sin(n\pi/2)}{n^2\pi^2} \tag{3.40}$$

Hence,

$$A_n = -\frac{8V_0 \sin n\pi/2}{n^2\pi^2 \sinh \frac{n\pi}{b}(a-w)} \tag{3.41}$$

It is instructive to find the capacitance per unit length $\mathcal{C}$ of the strip line using the fact that the energy stored per unit length is related to $\mathcal{C}$ according to

$$\mathscr{W} = \frac{1}{2}\mathcal{C}V^2 \tag{3.42}$$

where

$$\mathscr{W} = \frac{1}{2}\int \mathbf{D}\cdot\mathbf{E}\mathrm{d}S \tag{3.43}$$

For region 1,

$$\mathbf{E} = -\nabla V_1 = -\frac{\partial V_1}{\partial x}\mathbf{a_x} - \frac{\partial V_1}{\partial y}\mathbf{a_y} = \frac{2V_0}{b}\mathbf{a_y} \tag{3.44}$$

Hence, the energy per unit length in this region is

$$\mathscr{W}_1 = \frac{1}{2}\epsilon\int_{x=0}^{w/2}\int_{y=-b/2}^{0}\frac{4V_0^2}{b^2}\mathrm{d}y\mathrm{d}x = \frac{\epsilon w V_0^2}{2b} \tag{3.45}$$

For region 2,

$$E_x = -\frac{\partial V_2}{\partial x} = \sum_{n=1,3,5}^{\infty}\frac{n\pi}{b}A_n\cosh\left(\frac{n\pi}{b}(a/2-x)\right)\sin\frac{n\pi y}{b} \tag{3.46a}$$

$$E_y = -\frac{\partial V_2}{\partial y} = -\sum_{n=1,3,5}^{\infty}\frac{n\pi}{b}A_n\sinh\left(\frac{n\pi}{b}(a/2-x)\right)\cos\frac{n\pi y}{b} \tag{3.46b}$$

and

$$\begin{aligned}
\mathscr{W}_2 &= \frac{1}{2}\epsilon\int\int\left(E_x^2+E_y^2\right)\mathrm{d}x\mathrm{d}y\\
&= \frac{1}{2}\epsilon\int_{y=-b/2}^{0}\int_{x=w/2}^{0}\sum_n\sum_m\frac{mn\pi^2}{b^2}A_mA_n\\
&\left[\begin{matrix}\sinh\left(\frac{m\pi}{b}(a/2-x)\right)\sinh\left(\frac{n\pi}{b}(a/2-x)\right)\cos\frac{m\pi y}{b}\cos\frac{n\pi y}{b}\\ +\cosh\left(\frac{m\pi}{b}(a/2-x)\right)\cosh\left(\frac{n\pi}{b}(a/2-x)\right)\sin\frac{m\pi y}{b}\sin\frac{n\pi y}{b}\end{matrix}\right]\mathrm{d}x\mathrm{d}y
\end{aligned} \tag{3.47}$$

where the double summation is used to show that we are multiplying two series which may have different indices m and n. Due to the orthogonality property of sine and cosine functions, all terms vanish except when $m = n$. Thus,

$$\begin{aligned} \mathscr{W}_2 &= \frac{1}{2}\epsilon \sum_{n=odd}^{\infty} \frac{n^2\pi^2 A_n^2}{b^2} \frac{b}{2} \int_{w/2}^{a/2} \left\{ \sinh^2\left(\frac{n\pi}{b}(a/2 - x)\right) \right. \\ &\quad \left. + \cosh^2\left(\frac{n\pi}{b}(a/2 - x)\right) \right\} dx \\ &= \frac{1}{2}\epsilon \sum_{n=odd}^{\infty} \frac{n^2\pi^2 A_n^2}{4b} \frac{b}{n\pi} \cosh\left(\frac{n\pi}{2b}(a-w) \sinh\left(\frac{n\pi}{2b}(a-w)\right)\right) \end{aligned}$$

Substituting for A_n gives

$$\mathscr{W}_2 = \sum_{n=odd}^{\infty} \frac{8\epsilon V_0^2}{n^3\pi^3} \coth\left(\frac{n\pi}{2b}(a-w)\right) \tag{3.48}$$

The total energy in the four quadrants is

$$\mathscr{W} = 4\left(\mathscr{W}_1 + \mathscr{W}_2\right) \tag{3.49}$$

Thus,

$$\begin{aligned} \mathcal{C} = \frac{2\mathscr{W}}{V_0^2} &= \frac{8}{V_0^2}\left(\mathscr{W}_1 + \mathscr{W}_2\right) \\ &= \epsilon\left[\frac{4w}{b} + \frac{64}{\pi^3}\sum_{n=odd}^{\infty}\frac{1}{n^3}\coth\left(\frac{n\pi}{b}(a-w)\right)\right] \end{aligned} \tag{3.50}$$

The characteristic impedance of the lossless transmission line is given by

$$Z_0 = \frac{\sqrt{\mu\epsilon}}{\mathcal{C}} = \frac{\sqrt{\mu_r\epsilon_r}}{c\mathcal{C}}$$

or

$$Z_0 = \frac{120\pi}{\sqrt{\epsilon_r}\left[\frac{4w}{b} + \frac{64}{\pi^3}\sum_{n=odd}^{\infty}\frac{1}{n^3}\coth\left(\frac{n\pi}{b}(a-w)\right)\right]} \tag{3.51}$$

where $c = 3 \times 10^8$ m/s, the speed of light in a vacuum, and $\mu_r = 1$ is assumed.

3.5 Scattering by a Conducting Cylinder

The scattering of plane waves by a conductor structure is a very important problem. Many practical objects such as airplanes and missiles are cylindrical scatterers. Also, scattering by a conducting cylinder is one of the simplest cases with exact, analytical solution [4,5]. We will consider two possible cases: transverse magnetic $(\text{TM})^z$ and transverse electric $(\text{TE})^z$.

Case A—TM to z: Consider a plane wave traveling in the positive x-direction and polarized along z; that is, $\mathbf{E} = E_0 e^{-jkx}\mathbf{a_x}$, with $e^{j\omega t}$ time dependence assumed. Our objective is to determine the scattered field. Since the cylinder is infinitely long, the problem is reduced to two-dimensional as shown in Figure 3.3.

It is necessary to express the plane waves by cylindrical wave function using series expansion. We let

$$e^{-jx} = e^{-j\rho\cos\varphi} = \sum_{n=-\infty}^{\infty} a_n J_n(\rho)e^{jn\varphi} \tag{3.52}$$

where a_n is expansion coefficient to be determined. Since $e^{jn\varphi}$ for integer values of n represent a set of orthogonal functions, multiplying both sides of Equation 3.52 by $e^{jm\varphi}$ and integrating over $0 \leq \varphi \leq 2\pi$ gives

$$\int_0^{2\pi} e^{-j\rho\cos\varphi}e^{jm\varphi} = 2\pi a_m J_m(\rho) \tag{3.53}$$

Taking the mth derivative of both sides with respect to ρ and evaluating at $\rho = 0$ leads to

$$2\pi\frac{j^m}{2^m} = 2\pi a_m\frac{1}{2^m} \qquad \rightarrow \qquad a_m = j^{-m} \tag{3.54}$$

Substituting this into Equation 3.52, we obtain

$$e^{-jx} = \sum_{n=\infty}^{\infty} j^{-n}J_n(\rho)e^{jn\varphi} \tag{3.55}$$

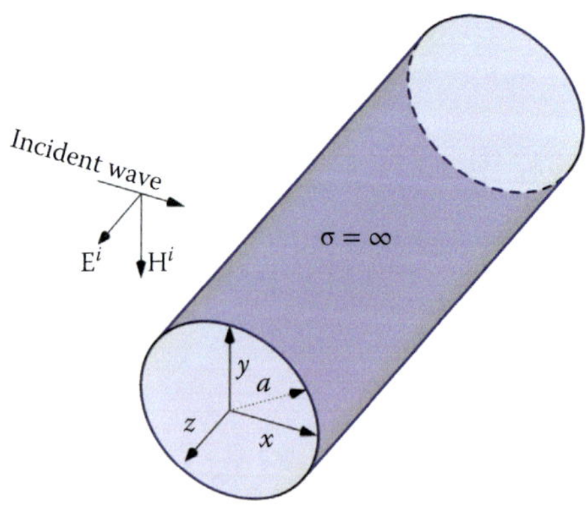

FIGURE 3.3
Incident uniform electromagnetic plane wave on a conducting cylinder (transverse magnetic to z case).

Thus, the incident wave may be written as

$$E_z^i = E_0 e^{-jx} = E_0 \sum_{n=\infty}^{\infty} j^{-n} J_n(k\rho) e^{jn\varphi} \tag{3.56}$$

Since the scattered field must consist of outgoing waves that vanish at infinity, it contains

$$J_n(k\rho) - jY_n(k\rho) = H_n^{(2)}(k\rho) \tag{3.57}$$

Hence,

$$E_z^s = \sum_{n=\infty}^{\infty} A_n H_n^{(2)}(k\rho) e^{jn\varphi} \tag{3.58}$$

where A_n is expansion coefficient. This expansion ensures that the correct radiation condition at infinity is met. The total field in medium 2 is

$$E_2 = E_z^i + E_z^s \tag{3.59}$$

while the total field in medium 1 is $E_1 = 0$ since medium 1 is conducting. At $\rho = a$, the boundary condition requires that the tangential components of E_1 and E_2 be equal. Hence,

$$E_z^i(\rho = a) + E_z^s(\rho = a) = 0 \tag{3.60}$$

Substituting Equations 3.56 and 3.58 into Equation 3.60,

$$\sum_{n=-\infty}^{\infty} \left[E_0 - (-j)^n J_n(ka) + A_n H_n^{(2)}(k\rho)\right] e^{jn\varphi} = 0 \tag{3.61}$$

From this, we obtain

$$A_n = -\frac{E_0(-j)^n J_n(ka)}{H_n^{(2)}(ka)} \tag{3.62}$$

Finally, substituting A_n into Equation 3.58 leads to the scattered wave as

$$\mathbf{E}_z^s = -E_0 \mathbf{a_z} \sum_{n=-\infty}^{\infty} (-j)^n \frac{J_n(ka) H_n^{(2)}(k\rho)}{H_n^{(2)}(ka)} \tag{3.63}$$

The corresponding scattered magnetic field components may be obtained by applying Maxwell's equations.

Case B—TE to z: Let us now assume that the incident wave is polarized transversely to z and is traveling along the $+x$ direction, as shown in Figure 3.4.

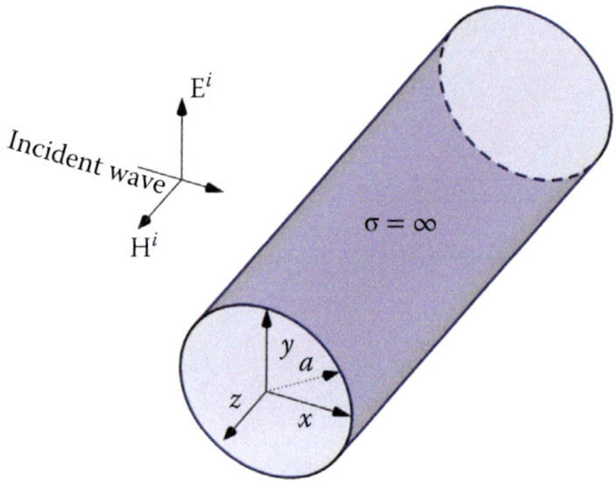

FIGURE 3.4
A uniform plane electromagnetic plane wave incident on a conducting cylinder.

It is incident normally on a perfectly conducting cylinder of radius a and of infinite length. The incident magnetic field is

$$H_z^i = H_0 \mathrm{e}^{-jkx} = H_0 \sum_{n=\infty}^{\infty} j^{-n} J_n(k\rho) \mathrm{e}^{jn\varphi} \tag{3.64}$$

The scattered magnetic field is

$$H_z^s = \sum_{n=\infty}^{\infty} B_n H_n^{(2)}(k\rho) \mathrm{e}^{jn\varphi} \tag{3.65}$$

where B_n are coefficients to be determined from the boundary condition. The boundary condition is $E_\varphi(\rho = a) = 0$. But

$$\begin{aligned} E_\varphi &= \frac{1}{j\omega\epsilon} \left[\nabla \times \left(H_z^i + H_z^s \right) \mathbf{a_z} \right]_\varphi \\ &= \frac{jk}{\omega\epsilon} \sum_{n=-\infty}^{\infty} \left[j^n H_0 J_n'(k\rho) + B_n H_n^{(2)}(k\rho) \right] \mathrm{e}^{jn\varphi} \end{aligned} \tag{3.66}$$

where the prime denotes differentiation. The boundary condition is satisfied when

$$B_n = -\frac{J_n'(ka)}{H_n^{(2)}(ka)} j^{-n} \tag{3.67}$$

Substituting this in Equation 3.65 gives the scattered field as

$$H_z^s = -\sum_{n=-\infty}^{\infty} j^{(-n)} \frac{J_n'(ka)}{H_n^{(2)}(ka)} \mathrm{e}^{jn\varphi} \tag{3.68}$$

Again, other field components may be obtained by using Maxwell's equations.

3.6 Scattering by a Dielectric Sphere

The scattering of EM waves by a dielectric sphere, known as the Mie scattering problem due to its first investigator in 1908, is an important problem whose exact analytical solution is usually referred to in assessing some numerical computations. Though the analysis of the problem is more rigorous, the procedure is similar to that of the previous section. Our treatment will be brief; for a more in-depth treatment, consult Stratton [6]. Consider a dielectric sphere illuminated by a plane wave propagating in the z direction and **E** is polarized in the x-direction as shown in Figure 3.5.

The incident wave is described as

$$\mathbf{E}^i = E_0 e^{j(\omega t - kz)} \mathbf{a_x} \tag{3.69a}$$

$$\mathbf{H}^i = \frac{E_0}{\eta} e^{j(\omega t - kz)} \mathbf{a_y} \tag{3.69b}$$

The first step is to series-expand this incident wave in terms of spherical wave functions. Since $e^{jz} = e^{jr\cos\theta}$ is independent of φ and finite at the origin, we let

$$e^{jz} = e^{jr\cos\theta} = \sum_{n=0}^{\infty} a_n j_n(r) P_n(\cos\theta) \tag{3.70}$$

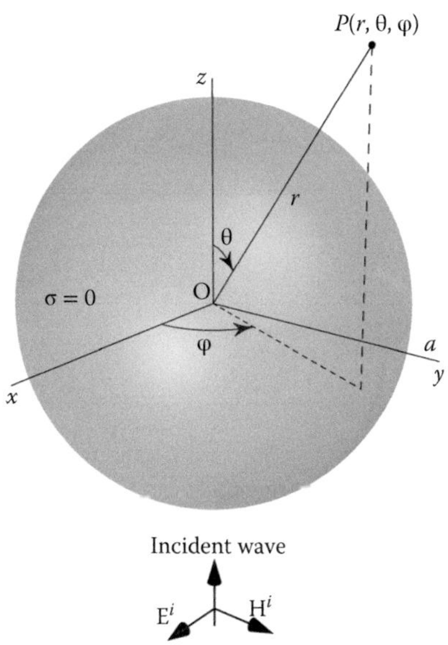

FIGURE 3.5
Scattering of a uniform plane wave incident on a dielectric sphere.

To determine the expansion coefficients a_n, we multiply both sides of Equation 3.69 by $P_m(\cos\theta)\sin\theta$ and integrate over $0 < \theta < \pi$

$$\begin{aligned}\int_0^{\pi} e^{jr\cos\theta} P_m(\cos\theta)\sin\theta d\theta &= \sum_{n=0}^{\infty} a_n j_n(r) \int_{-1}^{1} P_n(x)P_m(x)dx \\ &= \begin{cases} 0, & n \neq m \\ \frac{2}{2n+1} a_n j_n(r) & n = m \end{cases} \end{aligned} \tag{3.71}$$

where the orthogonality property has been applied. Taking the nth derivative of both sides and evaluating at $r = 0$ gives

$$j^n \int_0^{\pi} \cos^n \theta P_m(\cos\theta)\sin\theta d\theta = \frac{2}{2n+1} a_n \frac{d^n}{dr^n} j_n(r)\Big|_{r=0} \tag{3.72}$$

The left-hand side of Equation 3.72 yields

$$j^n \int_{-1}^{1} x^n P_n(x)dx = \frac{2^{n+1}(n!)^2}{(2n+1)!} j^n \tag{3.73}$$

To evaluate the right-hand side of Equation 3.72, we recall that

$$j_n(r) = \sqrt{\frac{\pi}{2r}} J_{n+1/2}(r) = \sqrt{\frac{\pi}{2}} \sum_{m=0}^{\infty} \frac{(-1)^m r^{2m+n}}{m!\Gamma(m+n+3/2)2^{2m+n+1/2}} \tag{3.74}$$

Hence,

$$\frac{d^n}{dr^n} j_n(r)\Big|_{r=0} = \sqrt{\frac{\pi}{2r}} \frac{n!}{\Gamma(n+3/2)2^{n+1/2}} = \frac{2^n(n!)^2}{(2n+1)!} \tag{3.75}$$

Substituting Equations 3.73 and 3.75 into Equation 3.72 gives

$$a_n = j^n(2n+1)$$

Thus,

$$e^{jz} = e^{jr\cos\theta} = \sum_{n=0}^{\infty} j^n(2n+1) j_n(r) P_n(\cos\theta) \tag{3.76}$$

or

$$E_x^i = E_0 e^{-jkz} = E_0 e^{-jkr\cos\theta} = E_0 \sum_{n=0}^{\infty} (-j)^n (2n+1) j_n(kr) P_n \cos\theta \tag{3.77}$$

Since,

$$\mathbf{a}_x = \sin\theta\cos\varphi\mathbf{a}_r + \cos\theta\cos\varphi\mathbf{a}_\theta - \sin\varphi\mathbf{a}_\varphi \tag{3.78}$$

The r-component of $\mathbf{E}^i$, for example, is

$$E_r^i = \cos\varphi\sin\theta E_i^r = E_0 e^{-j\omega t} \frac{\cos\varphi}{jkr} \frac{\partial}{\partial\theta} \left(e^{-jkr\cos\theta}\right) \tag{3.79}$$

Introducing this in Equation 3.77

$$E_r^i = E_0 e^{-j\omega t} \frac{\cos\varphi}{jkr} \sum_{n=0}^{\infty} (-j)^n (2n+1) j_n(kr) \frac{\partial}{\partial\theta} P_n(\cos\theta) \tag{3.80}$$

But

$$\frac{\partial P_n}{\partial\theta} = P_n^1$$

Hence,

$$E_r^i = E_0 e^{-j\omega t} \frac{\cos\varphi}{jkr} \sum_{n=1}^{\infty} (-j)^n (2n+1) j_n(kr) P_n^1(\cos\theta) \tag{3.81}$$

where the term $n = 0$ has been dropped since $P_0^1 = 0$. The same steps can be taken to express E_θ^i and E_φ^i in terms of the spherical wave functions. When those components are combined, the result is

$$\begin{aligned}\mathbf{E}^i &= \mathbf{a}_x E_0 e^{\omega t - kz} \\ &= E_0 e^{\omega t} \sum_{n=1}^{\infty} (-j)^n \frac{2n+1}{n(n+1)} \left[\mathbf{M}_n^{(1)}(k) + j\mathbf{N}_n^{(1)}(k)\right]\end{aligned} \tag{3.82}$$

$$\begin{aligned}\mathbf{H}^i &= \mathbf{a}_y E_0 e^{\omega t - kz} \\ &= -\frac{kE_0}{\mu\omega} e^{\omega t} \sum_{n=1}^{\infty} (-j)^n \frac{2n+1}{n(n+1)} \left[\mathbf{M}_n^{(1)}(k) - j\mathbf{N}_n^{(1)}(k)\right]\end{aligned} \tag{3.83}$$

where

$$\mathbf{M}_n^{(1)}(k) = \frac{1}{\sin\theta} z_n(kr) P_n^1(\cos\theta) \cos\varphi \mathbf{a}_\theta - z_n(kr) \frac{\partial P_n^1(\cos\theta)}{\partial\theta} (\sin\varphi) \sin\varphi \mathbf{a}_\varphi \tag{3.84}$$

$$\begin{aligned}\mathbf{N}_n^{(1)}(k) &= \frac{n(n+1)}{kr} z_n(kr) P_n^1(\cos\theta) \cos\varphi \mathbf{a}_r + \frac{1}{kr} \frac{\partial}{\partial r} [z_n(kr)] \frac{\partial P_n^1(\cos\theta)}{\partial\theta} \mathbf{a}_\theta \\ &+ \frac{1}{kr\sin\theta} \frac{\partial}{\partial r} [z_n(kr)] P_n^1(\cos\theta) \sin\varphi \mathbf{a}_\varphi\end{aligned} \tag{3.85}$$

The superscript (1) on the spherical vector functions **M** and **N** in Equations 3.82 and 3.83 indicates that these functions are constructed with spherical Bessel functions of the first kind; that is, $z_n(kr)$ in Equations 3.84 and 3.85 is replaced by $j_n(kr)$ when **M** and **N** are substituted in Equations 3.82 and 3.83. The induced secondary field consists of two parts. One part applies to the interior of the sphere and is referred to as the transmitted field, while the other applies to the exterior of the sphere and is called the scattered field. Thus, the total field outside the sphere is the sum of the incident and scattered fields. We now construct these fields in a fashion similar to the incident field. For the scattered field, we let

$$\mathbf{E}_s = E_0 e^{j\omega t} \sum_{n=1}^{\infty} (-j)^n \frac{2n+1}{n(n+1)} \left[a_n \mathbf{M}_n^{(4)}(k) + j b_n \mathbf{N}_n^{(4)}(k) \right] \quad (3.86)$$

$$\mathbf{H}_s = -\frac{kE_0}{\mu\omega} e^{j\omega t} \sum_{n=1}^{\infty} (-j)^n \frac{2n+1}{n(n+1)} \left[a_n \mathbf{M}_n^{(4)}(k) - j b_n \mathbf{N}_n^{(4)}(k) \right] \quad (3.87)$$

where a_n and b_n are expansion coefficients and the superscript (4) on $\mathbf{M}$ and $\mathbf{N}$ shows that these functions are constructed with spherical Bessel functions of the fourth kind (or Hankel function of the second kind); that is, $z_n(kr)$ in Equations 3.84 and 3.85 is replaced by $h_n^{(2)}(kr)$ when $\mathbf{M}$ and $\mathbf{N}$ are substituted in Equations 3.86 and 3.87. The spherical Hankel function has been chosen to satisfy the radiation condition. In other words, the asymptotic behavior of $h_n^{(2)}(kr)$, namely,

$$h_n^{(2)}(kr) = j^{n+1} \frac{e^{-kr}}{kr} \quad (3.88)$$

when combined with the time factor $e^{j\omega}$ represents an outgoing spherical wave. Similarly, the transmitted field inside the sphere can be constructed as

$$\mathbf{E}_t = E_0 e^{j\omega t} \sum_{n=1}^{\infty} (-j)^n \frac{2n+1}{n(n+1)} \left[c_n \mathbf{M}_n^{(1)}(k) + j d_n \mathbf{N}_n^{(1)}(k) \right] \quad (3.89)$$

$$\mathbf{H}_t = -\frac{kE_0}{\mu\omega} e^{j\omega t} \sum_{n=1}^{\infty} (-j)^n \frac{2n+1}{n(n+1)} \left[c_n \mathbf{M}_n^{(1)}(k) - j d_n \mathbf{N}_n^{(1)}(k) \right] \quad (3.90)$$

where c_n and d_n are expansion coefficients, k_1 is the propagation constant in the sphere. The functions $\mathbf{M}_n^{(1)}$ and $\mathbf{N}_n^{(1)}$ in Equations 3.89 and 3.90 are obtained by replacing $z_n(kr)$ in Equations 3.84 and 3.85 by $j_n(k_1 r)$; that is, $j_n(k_1 r)$ is the only solution in this case since the field must be finite at the origin, the center of the sphere.

The unknown expansion coefficients a_n, b_n, c_n, and d_n are determined by letting the fields satisfy the boundary conditions, namely, the continuity of the tangential components of the total electric and magnetic fields at the surface of the sphere. Thus, at $r = a$,

$$\mathbf{a}_r \times \left(\mathbf{E}^i + \mathbf{E}^s - \mathbf{E}^t \right) = \mathbf{0} \quad (3.91a)$$

$$\mathbf{a}_r \times \left(\mathbf{H}^i + \mathbf{H}^s - \mathbf{H}^t \right) = \mathbf{0} \quad (3.91b)$$

This is equivalent to

$$E_\theta^i + E_\theta^s = E_\theta^t, \qquad r = a \quad (3.92a)$$

$$E_\varphi^i + E_\varphi^s = E_\varphi^t, \qquad r = a \quad (3.92b)$$

$$H_\theta^i + H_\theta^s = H_\theta^t, \qquad r = a \quad (3.92c)$$

$$H_\varphi^i + H_\varphi^s = H_\varphi^t, \qquad r = a \quad (3.92d)$$

Substituting Equations 3.82, 3.83, 3.86, 3.87, 3.89, and 3.90 into Equation 3.92, multiplying the resulting equations by $\cos\varphi$ or $\sin\varphi$ and integrating over $0 \le \varphi \le 2\pi$, and then multiplying by $\frac{dP_m^1}{d\theta}$ and integrating over $0 \le \theta \le 2\pi$, we obtain

$$j_n(ka) + a_n h_n^{(2)}(ka) = c_n j_n(k_1 a) \quad (3.93a)$$

$$\mu_1 [ka j_n(ka)]' + a_n \mu_1 \left[ka h_n^{(2)}(ka)\right]' = c_n \mu [k_1 a j_n(k_1 a)]' \quad (3.93b)$$

$$\mu_1 j_n(ka) + b_n \mu_1 h_n^{(2)}(ka) = d_n \mu j_n(k_1 a) \quad (3.93c)$$

$$k [ka j_n(ka)]' + b_n k \left[ka h_n^{(2)}(ka)\right]' = d_n \mu [k_1 a j_n(k_1 a)]' \quad (3.93d)$$

Solving Equation 3.93a and 3.93b gives a_n and c_n, while solving Equations 3.93c and 3.93d gives b_n and d_n. Thus, for $\mu = \mu_0 = \mu_1$,

$$a_n = \frac{j_n(\alpha)[m\alpha j_n(m\alpha)]' - j_n(m\alpha)[\alpha j_n(\alpha)]'}{h_n^{(2)}(\alpha)[m\alpha j_n(m\alpha)]' - j_n(m\alpha)\left[\alpha h_n^{(2)}(\alpha)\right]'} \quad (3.94a)$$

$$b_n = \frac{j_n(m\alpha)[\alpha j_n(\alpha)]' - m^2 j_n(m\alpha)[\alpha j_n(\alpha)]'}{h_n^{(2)}(\alpha)[m\alpha j_n(m\alpha)]' - m^2 j_n(m\alpha)\left[\alpha h_n^{(2)}(\alpha)\right]'} \quad (3.94b)$$

$$c_n = \frac{j/\alpha}{h_n^{(2)}(\alpha)[m\alpha j_n(m\alpha)]' - j_n(m\alpha)\left[\alpha h_n^{(2)}(\alpha)\right]'} \quad (3.94c)$$

$$d_n = \frac{j/\alpha}{h_n^{(2)}(\alpha)[m\alpha j_n(m\alpha)]' - m^2 j_n(m\alpha)\left[\alpha h_n^{(2)}(\alpha)\right]'} \quad (3.94d)$$

where $\alpha = ka = 2\pi a/\lambda$ and $m = k_1/k$ is the refractive index of the dielectric, which may be real or complex depending on whether the dielectric is lossless or lossy. The primes at the square brackets indicate differentiation with respect to the argument of the Bessel function inside the brackets; that is, $[xz_n(x)]' = \frac{\partial}{\partial x}[xz_n(x)]$. To obtain Equations 3.94c and 3.94d, we have made use of the Wronskian relationship:

$$j_N(x)\left[x h_n^{(2)}(x)\right]' - h_n^{(2)}(x)[x j_n(x)]' = -j/x \quad (3.95)$$

If the dielectric is lossy and its surrounding medium is free space,

$$k_1^2 = \omega\mu_0(\omega\epsilon_1 - j\sigma), \qquad k^2 = \omega\mu_0\epsilon_0 \quad (3.96)$$

so that the (complex) refractive index m becomes

$$m = \frac{k_1}{k} = \sqrt{\epsilon_r} = \sqrt{\epsilon_{r1} - j\frac{\sigma}{\omega\epsilon_0}} = m' - jm'' \quad (3.97)$$

The problem of scattering by a conducting sphere can be obtained as a special case of the problem considered above. Since the EM fields must vanish inside the conducting sphere, the right-hand sides of Equations 3.91a, 3.91b, 3.92a, and 3.92d must be equal to zero so that $(c_n = 0 = d_n)$

$$a_n = -\frac{j_n(\alpha)}{h_n^{(2)}(\alpha)} \tag{3.98a}$$

$$b_n = -\frac{[\alpha j_n(\alpha)]'}{\left[\alpha h_n^{(2)}(\alpha)\right]'} \tag{3.98b}$$

Thus, we have completed the Mie solution; the field at any point inside or outside the sphere can now be determined.

3.7 Conclusion

We have considered ways of applying series expansion to practical EM problems. To illustrate how this powerful analytical technique is used, we have specifically applied it to the problems of Poisson's equation in a cube, Poisson's equation in a cylinder, strip transmission line, scattering by a conducting cylinder, and scattering by a dielectric sphere. Needless to say, there are several more EM applications of series expansion.

References

[1] T. Myint-U, *Partial Differential Equations of Mathematical Physics*, 2nd ed., New York: North-Holland, 1980, pp. 179–181.

[2] L. C. Andrews, *Elementary Partial Differential Equations with Boundary Value Problems*, Orlando, FL: Academic Press, 1986, pp. 171–172, 429–435.

[3] K. Asano, "Electrostatic potential and field in a cylindrical tank containing charged liquid," *Proc. IEEE*, vol. 124, no. 12, Dec. 1977, pp. 1277–1281.

[4] R. F. Harrington, *Time-Harmonic Electromagnetic Fields*, New York: McGraw-Hill, 1961, pp. 232–238.

[5] C. A. Balanis, *Advanced Engineering Electromagnetics*, New York: John Wiley & Sons, 1989, pp. 602–634.

[6] J. A. Stratton, *Electromagnetic Theory*, New York: McGraw-Hill, 1951, pp. 394–421.

Problems

1. In the interval $-1 < x < 1$, a pulse function is given by

$$f(x) = \begin{cases} 0 & \text{for} \quad -1 < x < a \\ 1 & \text{for} \quad a < x < 1 \end{cases}$$

Show that for all $-1 < x < 1$

$$f(x) = \tfrac{1}{2}(1-a) - \tfrac{1}{2}\sum_{n=1}^{m} \left[P_{n+1}(a) - P_{n-1}(a)\right] P_n(x).$$

2. Show that over the interval $-1 < x < 1$

$$\sqrt{1-x} = \frac{2\sqrt{2}}{3} P_0(x) - 2\sqrt{2}\sum_{n=1}^{\infty} \left[\frac{1}{(2n-1)(2n+3)}\right] P_n(x).$$

3. Consider a lossless transmission line of characteristic impedance Z_0 and length ℓ placed along the z-axis terminated in a resistive load R_L at $z = \ell$. Assume that the line is connected to a step voltage of V_g at $t = 0$. Also assume that the source has an internal resistance of R_g. Develop an expression for the voltage at a general point in space and time $v(z, t)$. Use the series method. You might take the aid of lattice or bounce diagrams.

4. Obtain the electrostatic potential due to a circular toroidal conductor held at V_0. Let R be the mean radius and a be the radius of cross section.

5. Consider two identical spheres each of radius a but placed a distance 2ℓ apart. Suppose the spheres are held at $\pm V$ volts, obtain an expression for the potential at a general point between the spheres. Use the series method.

6. A point charge Q exists at $(0, 0, h)$. Let a perfectly conducting disc described by $x^2 + y^2 \le a^2, z = 0$ be placed below the charge. Find the charge induced on the disc using the series method. Extend your answer for the special case $a \to \infty$. Tally your result via method of images.

4

Conformal Transformation

> There are two things to aim at in life: first, to get what you want; and, after that, to enjoy it. Only wisest of mankind achieve the second.
>
> – Logan P. Smith

CONTENTS

In the preceding two chapters, we have seen how an electromagnetics (EM) problem can be solved using variable separation technique and also by series expansion technique. In this chapter, we introduce another technique that helps in knowing potential and flux distributions in a broad class of two-dimensional problems. Also it can be used to characterize uniform transmission lines. The propagation of EM waves along a transmission line can be understood in terms of the primary constants of the line. If we assume that the line is lossless, the capacitance per unit length of the line helps us in obtaining the characteristic impedance of the line. This important parameter is related to the cross-sectional geometry of the line and the permittivity of the dielectric used.

To begin with, we introduce the fundamental notions of complex variables and the concept of analytic function and discuss several basic mappings [1–5].

The real and imaginary parts of an analytic function obey Cauchy–Riemann equations and individually satisfy the two-dimensional Laplace equation. An immediate consequence of this is the possibility of defining a complex potential function with real and imaginary parts denoting the flux function and the associated equipotential function.

4.1 Complex Variables

Complex numbers arise in the solution of polynomial equations. A complex number z is represented as a combination of two real numbers x and y with the help of the imaginary unit $j = \sqrt{-1}$. In Cartesian coordinates it is written as $z = x + jy$. The quantities x and y are called real and imaginary parts of z. We denote this fact as $x = \mathrm{Re}\{z\}, y = \mathrm{Im}\{z\}$ or $x = \Re\{z\}, y = \Im\{z\}$. In polar coordinates, it is represented as $z = \rho e^{j\varphi} = \rho\angle\varphi$; where ρ is the *modulus* or *amplitude* of z and φ is the *phase* or *argument* of z. As will be obvious, φ can take infinitely many values differing by integral multiples of 2π; the set of all these values is denoted by $\arg z$. However, the *principal value* of $\arg z$ is denoted by $\mathrm{Arg}\, z$ such that $-\pi < \mathrm{Arg}\, z \leq \pi$. Thus, $\rho = |z|$ and $\varphi = \arg z$. Using Euler's identity z can be written as $z = \rho(\cos\varphi + j\sin\varphi)$; which helps us to relate the two types of coordinates by

$$x = \rho\cos\varphi \qquad y = \rho\sin\varphi \tag{4.1}$$

$$\rho = \sqrt{x^2 + y^2}, \qquad \varphi = \arctan\frac{y}{x} \tag{4.2}$$

Note that while Cartesian form is defined all over the z-plane, the polar form is defined everywhere except at $z = 0$. Complex numbers are graphically represented in the xy-plane (called Argand plane) either in Cartesian form or in polar form as shown in Figure 4.1.

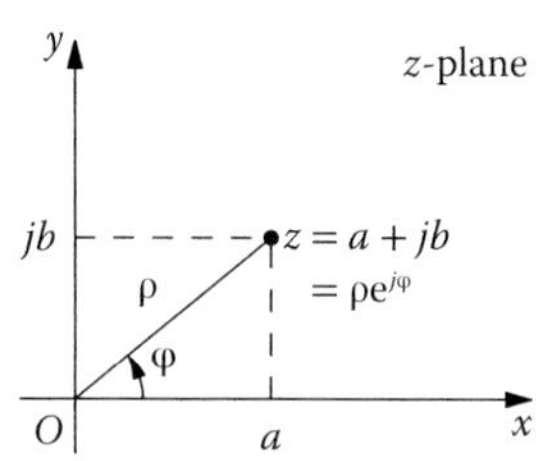

FIGURE 4.1
Cartesian and polar representation of a complex number in a plane.

The rules of arithmetic of complex numbers are similar to those of real number system. Thus, if $z_1 = x_1 + jy_1$ and $z_2 = x_2 + jy_2$,

$$z_1 \pm z_2 = (x_1 + jy_1) \pm (x_2 + jy_2) = (x_1 \pm x_2) + j(y_1 \pm y_2) \tag{4.3a}$$

$$z_1 z_2 = (x_1 + jy_1)(x_2 + jy_2) = (x_1x_2 - y_1y_2) + j(x_1y_2 + x_2y_1) \tag{4.3b}$$

$$\frac{z_1}{z_2} = \frac{(x_1 + jy_1)}{(x_2 + jy_2)} = \frac{(x_1x_2 + y_1y_2)}{x_2^2 + y_2^2} + j\frac{(x_2y_1 - x_1y_2)}{x_2^2 + y_2^2} \tag{4.3c}$$

the last holding true provided $|z_2|^2 = x_2^2 + y_2^2 \neq 0^2$.

In polar version, if $z_1 = \rho_1 e^{j\varphi_1}$ and $z_2 = \rho_2 e^{j\varphi_2}$,

$$\begin{aligned} z_1 \pm z_2 &= \rho_1 e^{j\varphi_1} \pm \rho_2 e^{j\varphi_2} \\ &= (\rho_1 \cos\varphi_1 \pm \rho_2 \cos\varphi_2) + j(\rho_1 \sin\varphi_1 \pm \rho_2 \sin\varphi_2) \end{aligned} \tag{4.4a}$$

$$z_1 z_2 = \rho_1\rho_2\{\cos(\varphi_1 + \varphi_2) + j\sin(\varphi_1 + \varphi_2)\} \tag{4.4b}$$

$$\frac{z_1}{z_2} = \frac{\rho_1}{\rho_2}\{\cos(\varphi_1 - \varphi_2) + j\sin(\varphi_1 - \varphi_2)\} \tag{4.4c}$$

Once again, the last holding true provided $|z_2|^2 = \rho_2^2 \neq 0^2$. If n is a positive integer, the power of a complex number is understood with the help of Euler and de Moivre formula; thus,

$$z^n = \{\rho e^{j\varphi}\}^n = \{\rho(\cos\varphi + j\sin\varphi)\}^n = \rho^n(\cos n\varphi + j\sin n\varphi) \tag{4.5a}$$

$$\begin{aligned} \sqrt[n]{z} &= \{\rho\left(\cos\left(\varphi + 2k\pi\right) + j\sin\left(\varphi + 2k\pi\right)\right)\}^{\frac{1}{n}} \\ &= \rho^{\frac{1}{n}}\left\{\cos\left(\frac{\varphi + 2k\pi}{n}\right) + j\sin\left(\frac{\varphi + 2k\pi}{n}\right)\right\} \\ &\qquad (k = 0, 1, 2, \ldots,\ n-1) \end{aligned} \tag{4.5b}$$

The *conjugate* of a complex number $z = x + jy$, denoted by z^* is defined as $z^* = x - jy$. The following are true in case of conjugate:

$$x = \frac{1}{2}(z + z^*) \tag{4.6a}$$

$$y = \frac{1}{2j}(z - z^*) \tag{4.6b}$$

$$zz^* = |z|^2 \tag{4.6c}$$

$$(z_1 z_2)^* = z_1^* z_2^* \tag{4.6d}$$

$$\Re\{z_1 z_2^*\} = \overrightarrow{z_2} \cdot \overrightarrow{z_1} \tag{4.6e}$$

$$\Im\{z_1 z_2^*\} = |\overrightarrow{z_2} \times \overrightarrow{z_1}| \tag{4.6f}$$

where the symbol $\rightarrow$ treats the complex number as a radial vector or position vector drawn from the origin. The following inequalities are useful:

$$|z_1 \pm z_2| \leq |z_1| + |z_2| \tag{4.7a}$$

$$|z_1 \pm z_2| \geq \Big| \; |z_1| - |z_2| \; \Big| \tag{4.7b}$$

4.2 Functions of a Complex Variable

A function of a complex variable is defined in much the same way as a real function. Let D be a set of complex numbers. A *function* f defined on D is a rule that assigns to each $z \in D$ a complex number w. The number w is called the image of z under f or simply the value of f at z and is denoted by $w = f(z) = u + jv$. The set D is called the *domain of definition* of f. The values of z are graphically shown in the usual xy-plane or the so called z-plane; whereas, those of w are marked in the uv- plane or the w-plane. If a given function involves simple operations like addition, subtraction, multiplication, and division it is easy to express it as $w = f(z) = u(x,y) + jv(x,y)$. But if its definition involves transcendental functions or if it is described by derivatives and integrals and so on, it is sometimes difficult to understand the mapping between the z-plane and the w-plane. Let us illustrate this idea with some simple examples.

Example 4.1. The electrostatic potential at a point (x,y) (in volts) is given by $\phi(x,y) = \mathfrak{Re}\left\{\frac{-j30}{\pi}\log z\right\}$. (i) Find its value at (5,1). (ii) Due to a spatial translation of electrodes it is modified as $\phi(x,y) = \mathfrak{Re}\left\{\frac{-j30}{\pi}\log(z-4)\right\}$. Repeat the problem.

Solution: Straightforward substitution gives the answers: (i) $\phi(5,1) = 1.885$ V (ii) $\phi(5,1) = 7.5$ V. We will recall this example when we discuss the Dirichlet's problem for coplanar electrodes in Section 4.6. □

Example 4.2. A complex variable z is subjected to the transformation $w = f(z) = z^2 + 5z + 41$. Obtain the real and imaginary parts $u(x,y)$ and $v(x,y)$ of $f(z)$. Also, given the points $A = (4 + j\,3)$, $B = (6 + j\,5)$, $C = (4.7 + j)$, $D = (5.3 + j\,7)$, and $P = (5 + j\,4)$ in the z-plane, find their images A', B', C', D', P', in the w-plane.

Solution: Substituting $z = x + jy$ in the given function we can write it as

$$\begin{aligned} w &= (x + jy)^2 + 5(x + jy) + 41 \\ &= (x^2 - y^2 + 5x + 41) + j(2xy + 5y) \end{aligned}$$

Thus,

$$u(x, y) = (x^2 - y^2 + 5x + 41)$$
$$v(x, y) = (2xy + 5y)$$

Substituting numerical values for the given points one obtains $A' = (68 + 39\jmath)$, $B' = (82 + 85\jmath)$, $C' = (85.59 + 14.4\jmath)$, $D' = (46.59 + 109.2\jmath)$, $P' = (75 + 60\jmath)$. These are shown in Figure 4.2.

□

Example 4.3. Extending Example 4.2, suppose the images of straight line segments $\overline{AB}$ and $\overline{CD}$ in the z-plane are the curves denoted by $\widetilde{A'B'}$ and $\widetilde{C'D'}$ in the w-plane. Compute and plot them together in respective planes.

Solution: Taking several points on each segment that are closely spaced, and using either a graphing calculator or a computer one might generate Figure 4.2. □

Example 4.4. Discuss and sketch the mapping of the directed lines $x = a$ and $y = b$ under the transformation $w = \mathrm{e}^z$.

Solution: Substituting $z = x + jy$, we get $w = \mathrm{e}^x \mathrm{e}^{jy} = R\mathrm{e}^{j\vartheta}$ where $R = e^x$ and $\vartheta = y$. So if we fix x as in the case of a vertical segment $x = a$, the image would be a circular arc of radius e^a. As y goes from 0 to 2π, we get a full circle. When x ranges from $-\infty$ to 0, the radius of the circle varies from 0 to unity. For $0 < x < \infty$, the radius of the circle goes from 1 to ∞ enclosing the whole w-plane. Similarly, if we fix $y = b$ the corresponding image is a radial line at an angle of $\vartheta = b$. This description is conveyed in Figure 4.3.

□

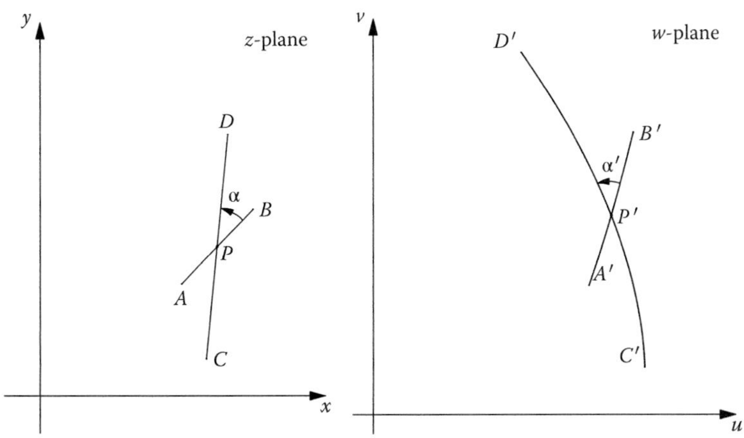

FIGURE 4.2
Images of $\overline{AB}$ and $\overline{CD}$ under the mapping $w = z^2 + 5z + 41$.

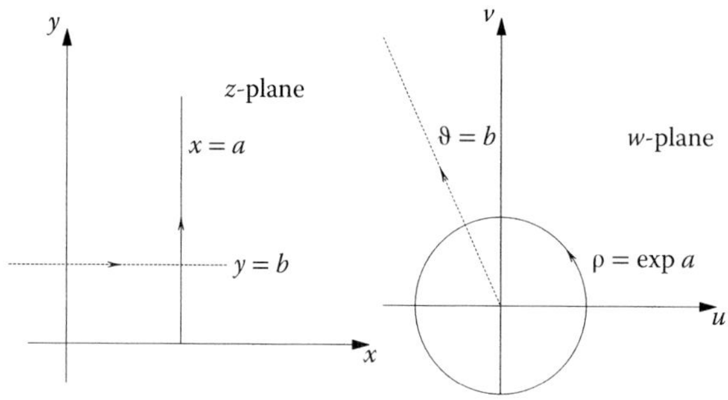

FIGURE 4.3
Mapping of $x = a$ (solid) and $y = b$ (dashed) under $w = \exp z$.

4.3 Derivative of a Complex Function

The idea of limits and continuity in real variable calculus extends to complex functions. If x and y approach x_0 and y_0, respectively, the complex variable z approaches $z_0 = x_0 + jy_0$. This helps us to extend the notion of continuity of functions. The function $f(z)$ is said to be continuous at the point z_0 if $f(z) \to f(z_0)$ whenever $z \to z_0$. Since $f(z) = u(x,y) + jv(x,y)$ and $f(z_0) = u(x_0, y_0) + jv(x_0, y_0)$, $f(z)$ is continuous at z_0 if $u(x,y)$ and $v(x,y)$ are both continuous at (x_0+jy_0). If $f(z)$ is continuous at every point in region R, it is said to be continuous in R. If the variable $z = x + jy$ changes by a small increment $\Delta z = \Delta x + j\Delta y$, it corresponds to a change Δw in $w = f(z)$. We define the derivative of the function $f(z)$ as

$$f'(z) = \frac{\mathrm{d}w}{\mathrm{d}z} = \lim_{\Delta z \to 0} \frac{f(z + \Delta z) - f(z)}{\Delta z} \tag{4.8}$$

provided the limit exists and is independent of the manner in which $\Delta z \to 0$. This requirement "shortlists" the classes of functions that do have a valid derivative; at the same time imparting several beautiful properties to them. Consider Figure 4.4 where a point $Q(z + \Delta z)$ attempts to approach $P(z)$. If the derivative $f'(z)$ should have a unique value, it should not matter how Q reaches P. Suppose Q reaches P along the horizontal line such that $\Delta y = 0, \ \Delta z = \Delta x$,

$$\frac{\mathrm{d}f}{\mathrm{d}z} = \frac{\partial f}{\partial x} = \frac{\partial u}{\partial x} + j\frac{\partial v}{\partial x} \tag{4.9}$$

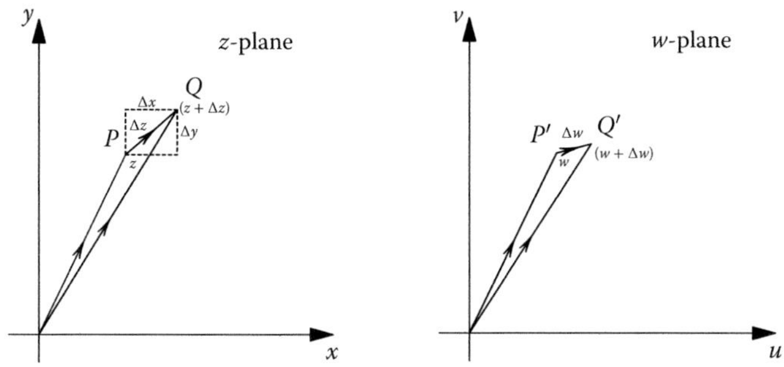

FIGURE 4.4
Differential increments in z- and w- planes.

Instead, if Q arrives along a vertical line such that $\Delta x = 0,\ \ \Delta z = j\Delta y$,

$$\begin{aligned}\frac{\mathrm{d}f}{\mathrm{d}z} &= \frac{\partial f}{j\,\partial y}\\ &= \frac{1}{j}\left(\frac{\partial u}{\partial x} + j\frac{\partial v}{\partial x}\right)\\ &= \frac{\partial v}{\partial y} - j\frac{\partial u}{\partial y}\end{aligned} \tag{4.10}$$

Since the derivative is assumed to be unique, Equations 4.9 and 4.10 must match. This requires that the functions $u(x, y)$ and $v(x, y)$ must satisfy

$$\frac{\partial u}{\partial x} = \frac{\partial v}{\partial y} \tag{4.11a}$$

$$\frac{\partial v}{\partial x} = -\frac{\partial u}{\partial y} \tag{4.11b}$$

These are known as the *Cauchy–Riemann equations* (CREs). Also, we know from basic calculus that the slope of a curve represented by $u(x, y) =$ constant at an arbitrary point is given by $\frac{dy}{dx} = \frac{u_x}{u_y}$ where $u_x = \frac{\partial u}{\partial x}$ and $u_y = \frac{\partial u}{\partial y}$. Then forming the product of slopes of $u(x, y) =$ constant and $v(x, y) =$ constant at a point we get $\frac{u_x v_x}{u_y v_y}$ which using CREs equals negative unity. That shows that the families of $u(x, y) =$ constant and $v(x, y) =$ constant curves are orthogonal.

Functions of complex variables that possess derivatives are called *analytic* or *holomorphic*. From CREs we find that if $u(x, y)$ is known $v(x, y)$ can be found up to an arbitrary constant and vice versa. Each of these real functions is said to be *conjugate* of the other. Differentiating Equation 4.11a with respect to x and Equation 4.11b with respect to y gives

$$\frac{\partial}{\partial x}\frac{\partial u}{\partial x} = \frac{\partial}{\partial x}\frac{\partial v}{\partial y} \tag{4.12a}$$

$$\frac{\partial}{\partial y}\frac{\partial v}{\partial x} = -\frac{\partial}{\partial y}\frac{\partial u}{\partial y} \tag{4.12b}$$

Adding

$$\frac{\partial^2 u}{\partial x^2} + \frac{\partial^2 u}{\partial y^2} = 0 \tag{4.13}$$

Similarly, differentiating Equation 4.11a with respect to y and Equation 4.11b with respect to x gives

$$\frac{\partial}{\partial y}\frac{\partial u}{\partial x} = \frac{\partial}{\partial y}\frac{\partial v}{\partial y} \tag{4.14a}$$

$$\frac{\partial}{\partial x}\frac{\partial v}{\partial x} = -\frac{\partial}{\partial x}\frac{\partial u}{\partial y} \tag{4.14b}$$

Adding

$$\frac{\partial^2 v}{\partial x^2} + \frac{\partial^2 v}{\partial y^2} = 0 \tag{4.15}$$

This concludes that the real and the imaginary parts of an analytic function satisfy the *Laplace's equation*. Using the symbol $\nabla^2 = \frac{\partial^2}{\partial x^2} + \frac{\partial^2}{\partial y^2}$ to denote the two-dimensional Laplacean operator,

$$\nabla^2 u(x, y) = 0 \tag{4.16a}$$

$$\nabla^2 v(x, y) = 0 \tag{4.16b}$$

To summarize, the following statements are equivalent in case of real and imaginary parts of an analytic function:

- The real and imaginary parts of an analytic function satisfy CREs.
- The families of curves represented by $u(x, y)$ = constant and $v(x, y)$ = constant curves are orthogonal.
- The real and imaginary parts of an analytic function are solutions of Laplace's equation.
- Each of the two functions $u(x, y)$ and $v(x, y)$ is said to be the *conjugate* of the other.

4.4 Conformal Mapping

Let f be an analytic function in the domain D, and let z_0 be a point in D. Let us also assume that $f'(z_0) \neq 0$. Writing $f(z)$ in the Taylor's series around the point z_0 as

$$\begin{aligned} f(z) &= f(z_0) + f'(z_0)(z - z_0) + f''(z_0)\frac{(z - z_0)^2}{2!} + \dots \\ &= f(z_0) + f'(z_0)(z - z_0) + \left[(z - z_0)\{f''(z_0)\frac{(z - z_0)}{2!} + \dots\}\right] \end{aligned} \tag{4.17}$$

and denoting the function in the bracket by $R(z)$, we can write

$$f(z) = f(z_0) + f'(z_0)(z - z_0) + R(z) \tag{4.18}$$

where $R(z) \to 0$ as $z \to z_0$. In particular, in the neighborhood of $z = z_0$, one might approximate Equation 4.18 as $f(z) \approx f(z_0) + f'(z_0)(z - z_0)$. This linear mapping is a combination of scaling, rotation, and translation. A curve $\{C : z(t) = x(t) + j\,y(t)\}$ is mapped as $\{G : w(t) = u\,(x(t), y(t)) + j\,v(x(t), y(t))\}$. The scaling is by a factor of $|f'(z_0)|$ and the rotation is by an angle $\alpha = \arg f'(z_0)$ and translation is over to a point $\{f(z_0) - z_0\, f'(z_0)\}$. A mapping $w = f(z)$ is said to be *conformal* at z_0 if it preserves angles between any two directed curves including the sense of their direction. In Figure 4.2 earlier, the angle $\alpha = \alpha' \approx 39.3°$. It can be easily shown that if $f(z)$ is analytic in the domain D containing z_0 and if $f'(z_0) \neq 0$, it is conformal at z_0. But if $f'(z_0) = 0$, while $f(z)$ is not constant everywhere, then we say that z_0 is a *critical point* of $f(z)$. In the event $f'(z_0) = 0$, we check to see the higher derivatives. If the first $(k-1)$ derivatives are zero at $z = z_0$, that is, $f^{(p)}(z_0) = 0$, for $1 \leq p \leq (k-1)$, the mapping magnifies the angles at $z = z_0$ by the factor k. This is clear from the Taylor series of $f(z)$ around z_0

$$f(z) = f(z_0) + \frac{f^{(k)}(z_0)}{k!}(z - z_0)^k + \frac{f^{(k+1)}(z_0)}{(k+1)!}(z - z_0)^{(k+1)} + \cdots \tag{4.19}$$

which can further be written as

$$f(z) - f(z_0) = (z - z_0)^k R(z) \tag{4.20}$$

This shows that the angles at $f(z_0)$ in the w-plane are related to those at z_0 in the z-plane according to

$$\arg\,(f(z) - f(z_0)) = k\,\arg\,(z - z_0) + \arg R(z)$$

We shall consider a few illustrative examples.

Example 4.5. Consider the mapping of $w = f(z) = z^2$. Discuss the transformation of the following:

(a) the unit square $\{0 \le x \le 1,\ 0 \le y \le 1,\}$
(b) a sector $\{\rho = a,\ 0 \le \varphi \le \alpha\}$
(c) and also sketch in the z-plane the pre-image of a rectangular region in the w-plane.

Solution:

(a) Let $z = x + jy$ and $w = u + jv$. Substituting for z we get

$$\begin{aligned} w &= (x + jy)^2 \\ &= (x^2 - y^2) + j2xy \\ u &= (x^2 - y^2), \qquad v = 2xy \end{aligned}$$

When $x = 0$ or $y = 0$, we get $v = 0$. If $x =$ constant, or say $x = a$, the locus of (u, v) follows a parabolic trajectory. Figure 4.5 shows how a unit square $S : 0 \le x, y \le 1$ maps into a region in the upper half-plane under the transform $w = f(z) = z^2$. Observe that the function is analytic everywhere, but $f'(0) = 0$, and $f''(0) = 2$. Accordingly, the angles at the three corners of the figure are preserved, but the angle at A is doubled.

(b) Let $z = \rho \exp j\varphi$ and $w = R \exp j\vartheta$. Substituting we get

$$\begin{aligned} w &= \rho^2 \exp j2\varphi \\ R &= \rho^2, \qquad \vartheta = 2\varphi \end{aligned}$$

This shows a circle of radius a is mapped into a circle of radius a^2. A radial line at an angle $\varphi = \alpha$ is mapped into a radial line $\vartheta = 2\alpha$. Figure 4.6 shows how a sector is mapped into another sector.

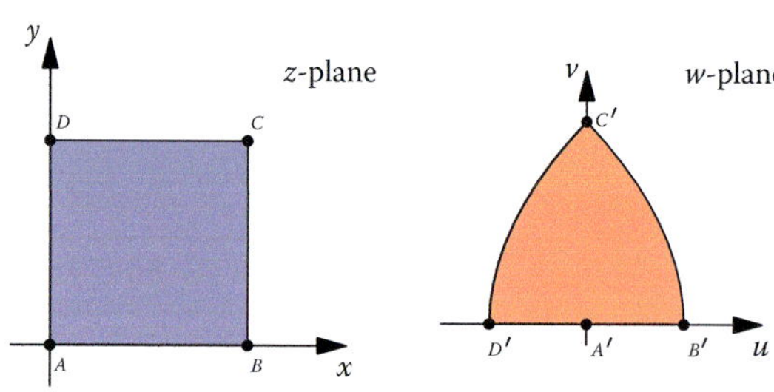

FIGURE 4.5
The unit square $0 \le x, y \le 1$ under $w = z^2$.

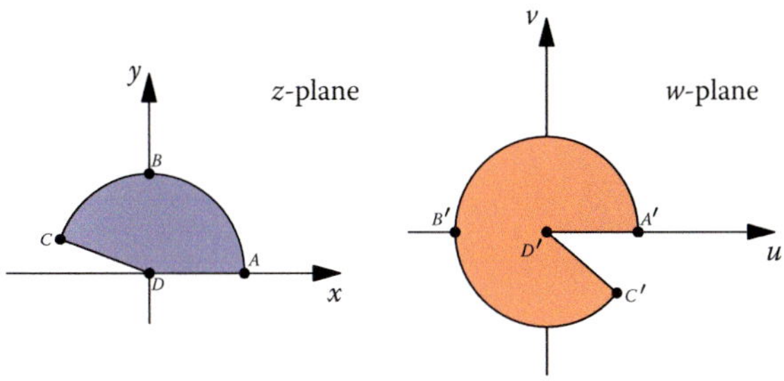

FIGURE 4.6
A sector of a unit circle under $w = z^2$.

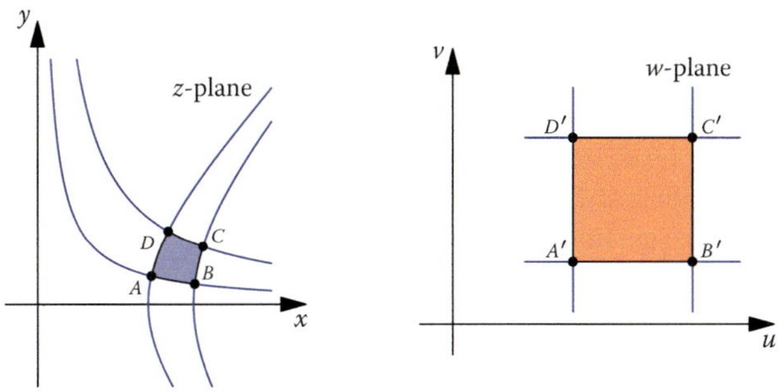

FIGURE 4.7
The preimage of a square under $w = z^2$.

(c) Likewise, if u = constant, say $u = a^2$ in the w-plane, this has a preimage of the hyperbola $x^2 - y^2 = a^2$ and the horizontal line $v = b$ has a preimage of $2xy = b$. Figure 4.7 shows how a region bounded by parabolas is mapped into a rectangle under $w = f(z) = z^2$.

□

Example 4.6. Continue Example 4.5 and sketch the image of a rectangle of height π under the mapping $w = f(z) = \exp(z)$.

Solution: As mentioned earlier, a vertical line $x = a$ is mapped into a circular arc of radius $\exp(a)$ and a horizontal line into a radial line of angle y radians. Thus for a given values of x, as y varies over an interval of length π, we get a semicircle in the upper half of w-plane. Thus, the given rectangle is transformed into a semiannular region as shown in Figure 4.8.

□

Example 4.7. Show that the reciprocal function $w = f(z) = \frac{1}{z}$ maps circles/lines into circles/lines.

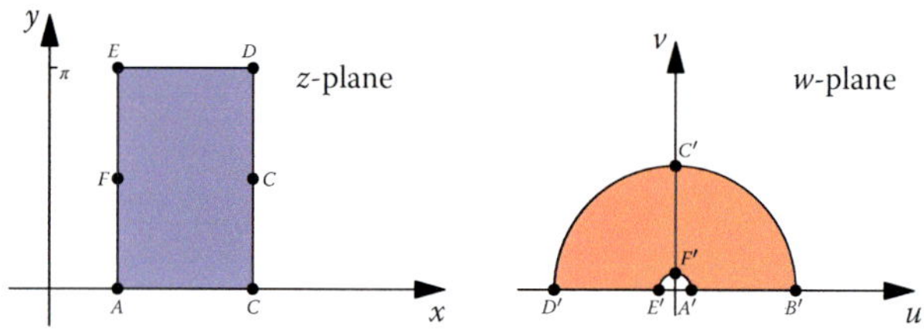

FIGURE 4.8
Mapping of a rectangle under $w = \exp z$.

Solution: The equation of a circle with the center $(-\alpha, -\beta)$ and radius $\sqrt{\alpha^2 + \beta^2 - \gamma}$ in the z-plane is given by $c(x^2+y^2)+2\alpha x+2\beta y+\gamma = 0$ where $c = 1$. Using Equation 4.6 this can be written as $czz^* + \alpha(z + z^*) - j\beta(z - z^*) + \gamma = 0$. We might view this to be in the form $czz^* + Az + Bz^* + \gamma = 0$. Substituting $z = 1/w$, we get $c + Aw^* + Bw + \gamma ww^* = 0$ which retains the same form. The case of $c = 0$ applies to straight lines. Thus, the reciprocal mapping maps circles/lines into circles/lines. □

Example 4.8. A mapping of the general form $w = f(z) = \frac{az+b}{cz+d}$, $(ad-bc \neq 0)$ is called the *bilinear transform or Mobius transform.*

(a) Show that under this mapping circles and/or straight lines are mapped as circles and/or straight lines. Consider a straight line as a special case of circle.
(b) Show that the inverse mapping is also bilinear.
(c) Show that there exists a *unique* mapping function that maps three given points z_1, z_2, and z_3 into w_1, w_2, and w_3, respectively.
(d) Two points P and $\tilde{P}$ are said to be symmetric with respect to the circle $C(o, r)$ whose center is o and radius is r provided o, P and $\tilde{P}$ are collinear and $\overline{oP} \cdot \overline{o\tilde{P}} = r^2$. Show that under bilinear mapping P and $\tilde{P}$ are mapped as P' and $\tilde{P}'$ that are symmetric with respect to the image of $C(o, r)$ denoted by $C'(o', r')$.
(e) Consider the two eccentric circles $C_o(0, 1) : x^2 + y^2 = 1$ and $C_i(b, r) : \left(x - \frac{x_1+x_2}{2}\right)^2 + y^2 = \left(\frac{x_2-x_1}{2}\right)^2$ and the region between them. Suggest a value for a in order that the function $w = f(z) = \frac{z-a}{az-1}$ maps these circles into a pair of concentric circles $C_i'(0, 1) : x^2 + y^2 = 1$ and $C_o'(0, R_0) : x^2 + y^2 = R_0^2$.

Solution:

(a) We can rewrite the function w as $w = \frac{a}{c} + \frac{(bc-ad)}{c^2(z+d/c)}$. This expression may be viewed as a combination of three different mappings as

$$s = z + d/c, \qquad t = 1/s, \qquad w = a/c + \frac{(bc-ad)t}{c^2}$$

But the first and the third are linear mappings and they possess the property mentioned. In Example 4.7, we proved this property for the reciprocal function. Combining all these ideas, we can affirm that the family of circles and straight lines is mapped into a family of circles and straight lines.

(b) We can solve for z; thus,

$$\begin{aligned} w &= \frac{az+b}{cz+d} \\ wcz + wd &= az + b \\ z(cw-a) &= -dw + b \\ z &= \frac{-dw+b}{cw-a} \end{aligned}$$

Obviously this is bilinear in its form.

(c) Consider the map $w = f(z)$ constructed from

$$\frac{w-w_1}{w-w_2} \times \frac{w_3-w_2}{w_3-w_1} = \frac{z-z_1}{z-z_2} \times \frac{z_3-z_2}{z_3-z_1}$$

This is bilinear. Substituting $z = z_k$, $k = 1,2,3$ yields $w = w_k$, $k = 1,2,3$. In order to prove its uniqueness, assume that one finds $w = f_1(z)$ that can equally map the given points so that $w_k = f_1(z_k)$, $k = 1,2,3$. Let $z = g(w)$ be the inverse mapping. Consider $g(f_1(z_k))$, $k = 1,2,3$. Being bilinear, it can be expressed as $z_k = \frac{-dz_k+b}{cz_k-a}$. This results in $cz_k^2 + (d-a)z_k - b = 0$ with three roots z_k, $k = 1,2,3$, which is absurd. Hence, $w = f_1(z) = f(z)$.

(d) Let P and $\tilde{P}$ be symmetric with respect to $C(o,r)$. Let G be any circle passing through P and $\tilde{P}$. Suppose we draw a tangent $\overline{oM}$ onto G, from geometry we recall that $\overline{oP} \cdot \overline{o\tilde{P}} = \overline{oM}^2$. Since P and $\tilde{P}$ are symmetric with respect to $C(o,r)$, $\overline{oM}^2$ must equal r^2 or G cuts $C(o,r)$ orthogonally at M. Now under bilinear mapping circles go to circles, the image of G must be a circle say, G' which intersects $C'(o',r')$ orthogonally thanks to conformality of the mapping to preserve angles. Hence applying the same argument, we find that P' and $\tilde{P}'$ are symmetric with respect to $C'(o',r')$.

(e) Note that we are asked to use a bilinear map to transform the given circles. In the w-plane because the circles are concentric, we recognize that $w = 0$ and $w = \infty$ are symmetric with respect to every circle C' around the origin. We also find that $z = a$ and $z = 1/a$ are their preimages and they should be symmetric with respect to every

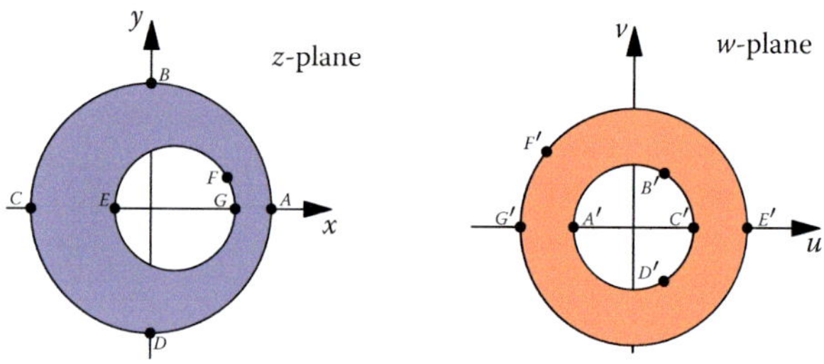

FIGURE 4.9
Bilinear transform mapping eccentric circles into concentric circles.

preimage c in the z-plane. If $z = a$ and $z = 1/a$ were to be symmetric with respect to C_i, it requires that

$$\left(a - \frac{x_1 + x_2}{2}\right)\left(1/a - \frac{x_1 + x_2}{2}\right) = \frac{x_1 - x_2}{4}$$

solving the quadratic for a and choosing the positive root, we get

$$a = \frac{1 + x_1 x_2 + \sqrt{(1 - x_1^2)(1 - x_2^2)}}{x_1 + x_2}$$

The mapping is shown in Figure 4.9. For some more simple examples on conformal mapping, the reader is suggested to consult various standard textbooks on complex variables, some of which are listed in References.

□

4.5 Complex Electric Potential

Suppose an electric field $\mathbf{E}$ is present in a dielectric of permittivity ϵ. In the three dimensional case, the flux crossing a differential area $\mathbf{dS} = dS\,\hat{\mathbf{n}}$ in the direction of the unit normal $\hat{\mathbf{n}}$ is given by $\epsilon\mathbf{E} \cdot \mathbf{dS}$. Also we recall from basic vector math that if the sides of an area are $\mathbf{dl_1}$ and $\mathbf{dl_1}$, their cross product has a magnitude equal to the area and the direction is in the direction of unit normal. Thus we can write flux as $\mathbf{E} \cdot \{\mathbf{dl_1} \times \mathbf{dl_2}\}$. This is shown in Figure 4.10.

We have found that the real and the imaginary parts of a complex analytic function satisfy the Laplace's equation and that they represent families of mutually orthogonal curves in the xy-plane. It is well-known that the equipotential lines and electric flux lines are orthogonal and that the potential satisfies

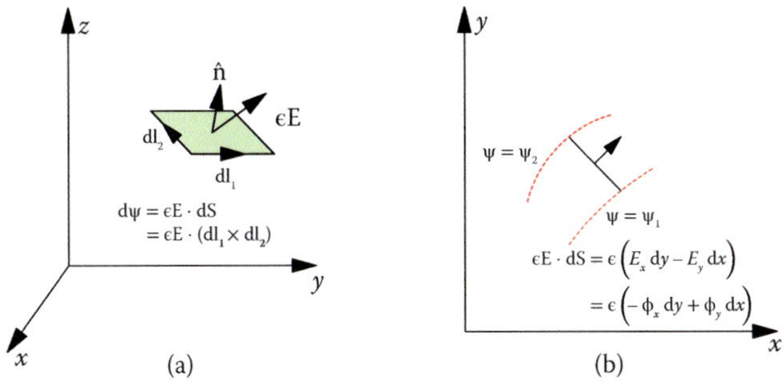

FIGURE 4.10
(a) Flux crossing a differential area **dS** in three dimensions. (b) Flux per unit length at a point from a reference in two dimensional case.

the Laplace's equation in a charge-free region. Then if we were to deal with electrostatic problems in two dimensions, say, in the xy-plane, with no variation in the z-direction, it is natural to conceive of a complex function having real and imaginary parts. If one of them represents the potential, then the other describes the flux function. Toward this end, let us assume that $\phi(x, y)$ stands for the electrostatic potential in a given region in the xy-plane. If there are two equipotential surfaces ϕ_1 and ϕ_2, the potential difference is given by

$$\text{Potential difference } = \int_{\phi_1}^{\phi_2} d\phi \tag{4.21}$$

$$\propto (\phi_2 - \phi_1) \tag{4.22}$$

Let us introduce the complex function $w = \Phi(z) = \phi(x, y) + j\psi(x, y)$ pending the interpretation of $\psi(x, y)$. The electric field vector associated with $\phi(x, y)$ is given by

$$\mathbf{E} = -\nabla\phi \tag{4.23a}$$

$$= -\frac{\partial\phi}{\partial x}\mathbf{a_x} - \frac{\partial\phi}{\partial y}\mathbf{a_y} \tag{4.23b}$$

But from CREs,

$$\frac{\partial\phi}{\partial x} = \frac{\partial\psi}{\partial y}, \quad \frac{\partial\psi}{\partial x} = -\frac{\partial\phi}{\partial y} \tag{4.24}$$

Hence,

$$\mathbf{E} = -\frac{\partial\psi}{\partial y}\mathbf{a_x} + \frac{\partial\psi}{\partial x}\mathbf{a_y} \tag{4.25}$$

If we take $\mathbf{dl_2}$ as a unit length along z-direction, that is, $\mathbf{dl_2} = 1\mathbf{a_z}$ and $\mathbf{dl_1} = \mathrm{d}x\mathbf{a_x} + \mathrm{d}y\mathbf{a_y}$, then flux per unit length crossing a differential length is given by

$$\begin{aligned}\epsilon\mathbf{E}\cdot\big((\mathrm{d}x\mathbf{a_x} + \mathrm{d}y\mathbf{a_y})\times\mathbf{a_z}\big) &= \epsilon\mathbf{E}\cdot(\mathrm{d}y\mathbf{a_x} - \mathrm{d}x\mathbf{a_y})\\ &= \epsilon\left(E_x\mathrm{d}y - E_y\mathrm{d}x\right)\\ &= \epsilon\left(-\frac{\partial\phi}{\partial x}\mathrm{d}y + \frac{\partial\phi}{\partial y}\mathrm{d}x\right)\\ &= \epsilon\left(-\frac{\partial\psi}{\partial y}\mathrm{d}y - \frac{\partial\psi}{\partial x}\mathrm{d}x\right)\\ &= -\epsilon\mathrm{d}\psi\end{aligned}$$

Then the net flux per meter length crossing an equipotential section between two references ψ_1 and ψ_2 (denoting it by Ψ_{21}) is given by

$$\begin{aligned}\text{Flux } \Psi_{21} &= \int_{\psi_1}^{\psi_2} -\epsilon\mathrm{d}\psi\\ &\propto (\psi_2 - \psi_1)\end{aligned}$$

This implies that just as potential difference is found by $\phi(x, y)$, the flux (per unit length in the z-direction) crossing an equipotential segment between two reference points is found by $\psi(x, y)$. Also, just as $\phi(x, y)$ alone yields potential *at a point* with an arbitrary choice of reference, even $\psi(x, y)$ can be interpreted as flux *at the same point* also with an arbitrary choice of reference. Strictly speaking, flux at a point makes no sense. First of all, a hard look at the CREs reveals that $\phi(x, y)$ and $\psi(x, y)$ have the same dimensions of volts. But in expressing the flux in terms of $\psi(x, y)$ we dropped the multiplication factor $-\epsilon$ whose units are farads/meter; and product of farads/meter with volts result in coul/m. Thus we get the units of flux per unit length. Thus, when we talk about flux at a point, we have at the back of our mind a reference, the need to multiply with $-\epsilon$ and a unit length in the direction of the z-axis.

An important application of this idea is in finding capacitance of transmission lines of various geometries. The capacitance per unit length $\mathscr{C}$ is defined as the ratio of charge per unit length to the applied potential between the conductors. From Gauss law we can also restate this as the ratio of flux per unit length to the potential difference between the conductors. Thus we can write

$$\mathscr{C} = \epsilon\left(\frac{\psi_1 - \psi_2}{\phi_1 - \phi_2}\right) \qquad \text{F/m} \tag{4.26}$$

The choice of taking the real part of $w = \Phi(z) = \phi(x, y) + j\psi(x, y)$ to represent potential is only arbitrary. Since $\phi(x, y)$ and $\psi(x, y)$ are conjugates of each other, one might choose the imaginary part for potential function; in which case the real part serves as flux function. Eventually, identifying the complex function is the crucial step in solving a given problem. We shall consider some examples.

Example 4.9. Construct a complex potential function $\Phi(z)$ to represent the problem of a line charge placed at the origin in the xy-plane (i.e., along the z-axis). Transform the same to represent the case of an electric double layer covering the positive xz-plane.

Solution: Consider the function

$$\begin{aligned}\Phi(z) &= -\frac{\rho_L}{2\pi\epsilon_0}\ln z \\ &= -\frac{\rho_L}{2\pi\epsilon_0}\ln\rho - j\frac{\rho_L}{2\pi\epsilon_0}\phi\end{aligned}$$

Clearly, we can identify the real part as the potential function of a line charge of density ρ_L while the imaginary part equated to a constant, that is, $\phi =$ constant representing flux lines.

Now suppose we simply multiply the preceding expression with j and again consider the complex potential we get

$$\begin{aligned}\Phi(z) &= -j\frac{\rho_L}{2\pi\epsilon_0}\ln z \\ &= \frac{\rho_L}{2\pi\epsilon_0}\phi - j\frac{\rho_L}{2\pi\epsilon_0}\ln\rho\end{aligned}$$

This time the real part represents the potential of a double layer $\rho_\pm$ covering the positive xz-plane and the imaginary part describes flux lines. These two cases are depicted in Figure 4.11a and b, respectively.

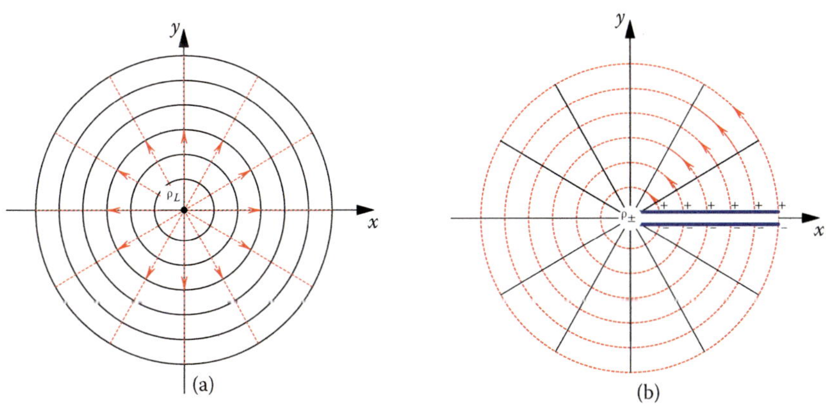

FIGURE 4.11
Equipotential and flux lines of (a) a line charge ρ_L along the z-axis and (b) a double layer $\rho_\pm$ covering the positive xz-plane.

It is interesting to find that the simple operation of multiplication with j has changed the problem altogether. This prompts us to think that other transformations could help us to solve more difficult electrostatic problems. The skill in coming up with the particular operation to perform is gained by intuition and some experience. In this chapter, we find it to be true as we work out a good number of examples. □

Example 4.10. Two infinite parallel plates $x = a$ and $x = b$, $b > a$ are held at $\phi = V_0$ and $\phi = 0$, respectively. Find the potential $\phi(x, y)$ in the region between them and suggest a complex potential function $\Phi(z)$.

Solution: For the given geometry, the potential function varies only with x. The equipotential lines are described families of parallel lines $x = C$. The Laplace's equation for ϕ becomes

$$\nabla^2\phi = \frac{\mathrm{d}^2\phi}{\mathrm{d}x^2} = 0$$

A possible solution is of the form

$$\phi(x, y) = Ax + B$$

Using the boundary conditions $\phi(x = a, y) = 0$ and $\phi(x = b, y) = V_0$ we can solve for A and B, and express the potential as

$$\phi(x, y) = V_0\left(\frac{x - b}{a - b}\right)$$

A complex potential function of the form $\Phi(z) = V_0\left(\frac{z-b}{a-b}\right)$ can be used to describe this problem with $\mathfrak{Re}\{\Phi(z)\} = \phi(x, y)$ to represent the potential. □

Example 4.11. Two coaxial cylinders $\rho = a$ and $\rho = b$, $b > a$, are held at $\phi = V_0$ and $\phi = 0$, respectively. Find the potential $\phi(x, y)$ in the region between them and suggest a complex potential function $\Phi(z)$.

Solution: For the given geometry, the potential function varies only with $|z| = \rho$. The equipotential lines are described by families of concentric circles $\rho = C$ in the xy-plane. The Laplace's equation for $\phi(\rho)$ becomes

$$\nabla^2\phi = \frac{1}{\rho}\frac{\mathrm{d}}{\mathrm{d}\rho}\left(\rho\,\frac{\mathrm{d}\phi}{\mathrm{d}\rho}\right) = 0$$

A possible solution is of the form

$$\phi(x, y) = A\ln\rho + B$$

Using the boundary conditions $\phi(\rho = a, \phi) = V_0$ and $\phi(\rho = b, \phi) = 0$, we get

$$V_0 = A\ln a + B$$
$$0 = A\ln b + B$$

we can solve for A and B, and express the potential as

$$\phi(x, y) = V_0 \left(\frac{\ln(\rho/b)}{\ln(a/b)} \right)$$

A complex potential function of the form $\Phi(z) = V_0 \left(\frac{\ln(z/b)}{\ln(a/b)} \right)$ can be used to describe this problem with $\Re e\{\Phi(z)\} = \phi(x, y)$ to represent the potential. □

Example 4.12. Discuss the mapping carried out by the analytic function $w = \ln z$. Solve for potential between the coaxial cylinders $\rho = a$ and $\rho = b$, $b > a$ held at $\phi = V_0$ and $\phi = 0$, respectively.

Solution: Consider the function $w = \log z$. Substituting $z = \rho e^{j\varphi}$, we get $w = \ln \rho + j\varphi$. This maps the annular region between the two cylinders described by $D = \{\rho e^{j\varphi} : a \le \rho \le b,\ 0 \le \varphi \le 2\pi\}$ into the rectangular strip $W = \{u + jv : \ln a < u < \ln b, 0 < v < 2\pi\}$. This is shown in Figure 4.12. The boundary condition in the w-plane is

$$w(\ln a, v) = V_0$$
$$w(\ln b, v) = 0$$

Now the geometry in the w-plane is similar to the parallel plate geometry in the z-plane of Example 4.10. Systematically changing symbols, the complex potential is given by

$$\Phi(w) = V_0 \left(\frac{w - \ln b}{\ln a - \ln b} \right)$$

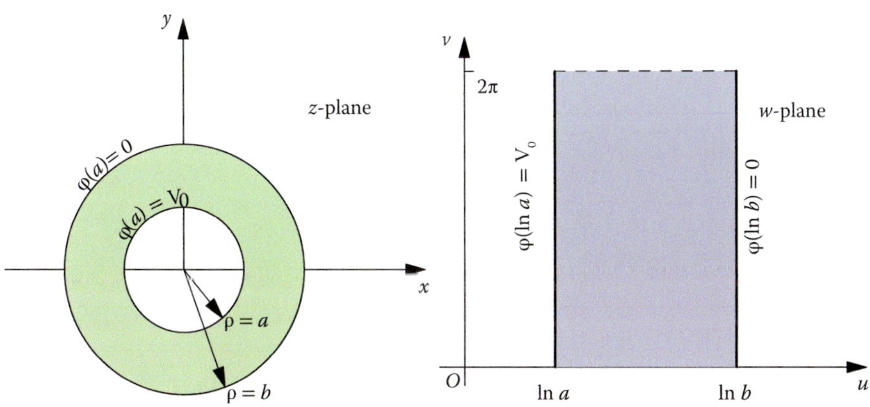

FIGURE 4.12
Mapping the coaxial geometry into parallel plate geometry.

Substituting for w, we can write

$$\begin{aligned}\Phi(z) &= V_0\left(\frac{\ln z - \ln b}{\ln a - \ln b}\right)\\ &= V_0\left(\frac{\ln(z/b)}{\ln(a/b)}\right)\end{aligned}$$

Taking the real part,

$$\begin{aligned}\phi(x,y) &= \mathfrak{Re}\left\{V_0\left(\frac{\ln(z/b)}{\ln(a/b)}\right)\right\}\\ &= V_0\left(\frac{\ln(\rho/b)}{\ln(a/b)}\right)\end{aligned}$$

□

Example 4.13. Two semi-infinite plates $\varphi = 0$ and $\varphi = \alpha$, $\rho \neq 0$, are held at $\phi = 0$ and $\phi = V_0$, respectively. Find the potential $\phi(x,y)$ in the region between them. Make use of your answer to find potential $\phi(x,y)$ in the upper half of the xy-plane if the part of x-axis to the right of $x = a$ is held at 0 V and the remaining part to the left of $x = a$ is held at V_0 with an infinitesimal gap at $x = a$.

Solution: For the given geometry, the potential function varies only with Arg $z = \varphi$. The equipotential lines are described by families of radial lines $\varphi = C$.

A possible solution is of the form

$$\phi(x,y) = A\varphi + B$$

Using the boundary conditions $\phi(\rho, \varphi = 0) = 0$ and $\phi(\rho, \varphi = \alpha) = V_0$, we get

$$\begin{aligned}0 &= A \times 0 + B\\ V_0 &= A\alpha + B\end{aligned}$$

we can solve for A and B, and express the potential as

$$\phi(x,y) = \left(\frac{V_0}{\alpha}\right)\varphi$$

A complex potential function of the form $\Phi(z) = \left(\frac{-jV_0}{\alpha}\right)\ln(z)$ can be used to describe this problem with $\mathfrak{Re}\{\Phi(z)\} = \phi(x,y)$ to represent the potential.

The modified geometry is obtained for the special case of $\alpha = \pi$ and the origin at $(a, 0)$. So the modified complex potential $\Phi(z)$ and the modified real potential $\phi(x, y)$ are given by

$$\Phi(z) = -j\left(\frac{V_0}{\pi}\right)\ln(z-a)$$

$$\begin{aligned}\phi(x,y) &= \mathfrak{Re}\{\Phi(z)\}\\ &= \left(\frac{V_0}{\pi}\right)\text{Arg}\{(z-a)\}\\ &= \left(\frac{V_0}{\pi}\right)\arctan\left(\frac{y}{x-a}\right)\end{aligned}$$

□

4.6 Coplanar Strips at Fixed Potentials

We just found that the potential in case of two half-planes at an angle of α held at V_0, is given by $\left(\frac{V_0\varphi}{\alpha}\right)$, and that if the planes are coplanar, it becomes $\left(\frac{V_0\varphi}{\pi}\right)$. Further, if the origin is shifted to a, we can express the potential as $\mathfrak{Re}\left\{-j\left(\frac{V_0}{\pi}\right)\ln(z-a)\right\}$. We are now in a position to extend these results to find the potential and flux functions in the upper half of the xy-plane due to several coplanar strips at fixed potentials.

Suppose the entire xz-plane is divided into N strips parallel to the z-axis and electrically isolated from each other by an infinitesimal gap and each is held at a fixed potential as shown in Figure 4.13a. This is the well-known *Dirichlet problem* for the upper half-plane. This may be generalized to a closed contour also. Instead of giving the proof in mathematical notation, let us use the idea of superposition that is very common in electrical engineering. Although various contiguous sections of the x-axis are held at different fixed potentials, we shall view this arrangement as a superposition of various subproblems shown in Figure 4.13b. Starting from the rightmost, we first assume that the whole x-axis is held at a potential of V_N. This sets a reference value. Next consider the pair of half-planes in the second row where $\phi = V_{N-1} - V_N$ for $x < x_{N-1}$ and 0 for $x > x_{N-1}$. A combination of these two results in a problem involving two half-planes separated by $x = x_{N-1}$ the left being held at V_{N-1} and the right half at V_N. Continuing thus, we can express the net complex potential, real potential, and the flux function as

$$\Phi(z) = \left\{V_N - \frac{j}{\pi}\sum_{k=1}^{N}\frac{V_k - V_{k+1}}{\pi}\ln(z-x_k)\right\} \tag{4.27}$$

$$\phi(x,y) = \mathfrak{Re}\{\Phi(z)\} \tag{4.28}$$

$$\psi(x,y) = \Im\{\Phi(z)\} \tag{4.29}$$

Next, we shall show some examples based on this idea.

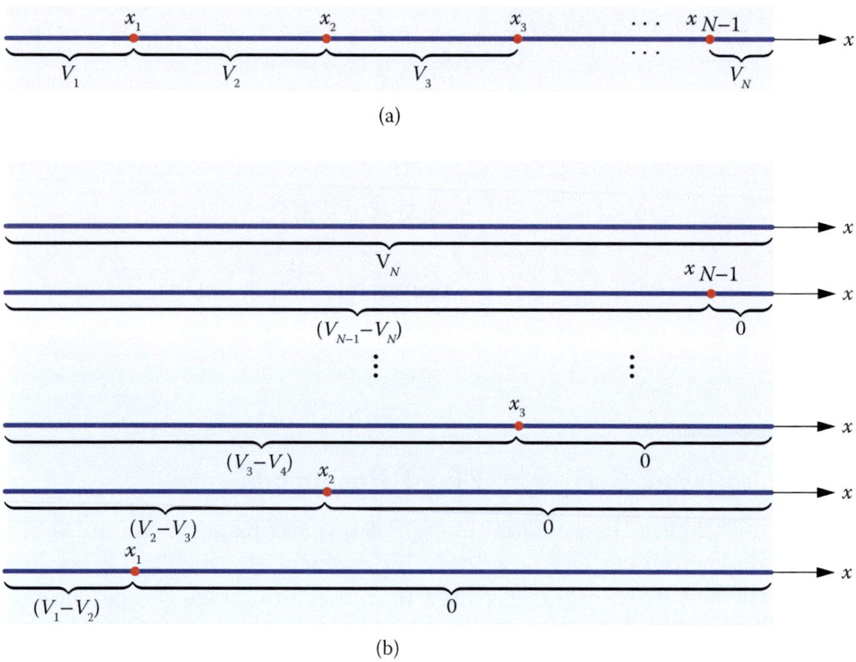

FIGURE 4.13
N coplanar strips held at different potentials. (a) Original problem (b) N separate problems.

Example 4.14. The infinite strip described by $\{y = 0,\ |x| < 1\}$ is held at 1 V while the remaining semi-infinite planes $\{y = 0,\ |x| > 1\}$ are grounded. Find the complex potential $\Phi(z)$ in the upper half of the xy-plane. From its real and imaginary parts sketch the equipotential and flux lines in the upper half of xy-plane.

Solution: We can set up the complex potential just by inspection. First, we write the reference potential on the right most strip. To this, we keep adding terms corresponding to every infinitesimal gap x_k. At each x_k, we take the potential on the left strip minus the potential on the right strip and include the term $\frac{-j}{\pi}\ln(z - x_k)$.

$$\begin{aligned}\Phi(z) &= 0 - \frac{j}{\pi}(0-1)\ln(z+1) - \frac{j}{\pi}(1-0)\ln(z-1) \\ &= \frac{j}{\pi}\ln\left(\frac{z+1}{z-1}\right)\end{aligned}$$

Taking real and imaginary parts

$$\phi(x, y) = \frac{1}{\pi} \arctan\left(\frac{2y}{x^2 + y^2 - 1}\right)$$

$$\psi(x, y) = \frac{1}{2\pi} \ln \frac{(x+1)^2 + y^2}{(x-1)^2 + y^2}$$

These are plotted in Figure 4.14. □

Example 4.15. The xz-plane has three infinitesimal gaps along $x = 0, \pm 1$. The strips are held at various potentials as follows:

$$\phi = \begin{cases} 4\,\text{V}, & \text{for} \quad x < -1 \\ 1\,\text{V}, & \text{for} \quad -1 < x < 0 \\ 3\,\text{V}, & \text{for} \quad 0 < x < 1 \\ 2\,\text{V}, & \text{for} \quad x > 1 \end{cases}$$

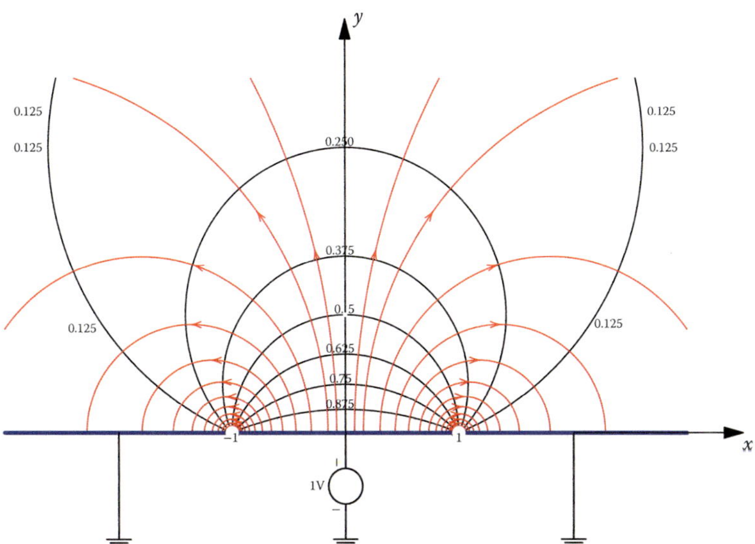

FIGURE 4.14
Three coplanar strips held at different potentials.

Find the complex potential $\Phi(z)$ in the upper half of the xy-plane. From its real and imaginary parts sketch the equipotential and flux lines in the upper half of the xy-plane.

Solution: The complex potential is given by

$$\begin{aligned}\Phi(z) &= 2 - \frac{j}{\pi}\left((4-1)\ln(z+1) + (1-3)\ln(z) + (3-2)\ln(z-1)\right)\\ &= 2 - \frac{j}{\pi}\left\{\ln(z-1)z^{-2}(z+1)^3\right\}\end{aligned}$$

Plotting the real and imaginary parts, we get Figure 4.15. □

Example 4.16. (a) An infinite cylinder of unit radius symmetric around the z-axis has been split into two halves between which a potential difference of 1 V is maintained. Consider its cross-section in the xy-plane and use bilinear mapping to transform it into two half-planes. Sketch the equipotential and flux lines in its cross section in the xy-plane. The potential on the cylinder may be written as

$$\phi = \begin{cases} 1\,\text{V}, & \text{for} \quad 0 < \varphi < \pi \\ 0\,\text{V}, & \text{for} \quad \pi < \varphi < 2\pi \end{cases}$$

(b) Modify the answer if only a quarter of the cylinder is split and a voltage of 80 V is maintained between the two parts as described by

$$\phi = \begin{cases} 80\,\text{V}, & \text{for} \quad 0 < \varphi < \pi/2 \\ 0\,\text{V}, & \text{for} \quad \pi/2 < \varphi < 2\pi \end{cases}$$

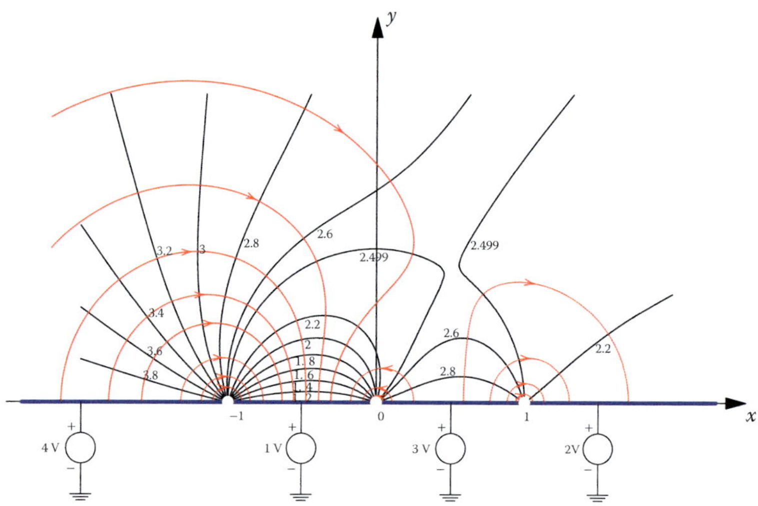

FIGURE 4.15
Four coplanar strips held at different potentials.

Solution: (a) Consider the function $w = j\frac{1-z}{1+z}$ which is bilinear. We know that it maps circles and straight lines into circles and straight lines; in particular, the unit circle in the z-plane is mapped into the upper half of the w-plane. It is easy to verify that the points $(1, 0)$, $(0, j)$, $(-1, 0)$, and $(0, -j)$ in the z-plane go respectively into $(0, 0)$, $(1, 0)$, $(\infty, 0)$, and $(-1, 0)$ in the w-plane. This mapping is shown in Figure 4.16. The geometry in the w-plane is rather similar to that of Example 4.13. We can write the complex potential function in the w-plane first and then by substitution for w get the answer in the z-plane.

$$\Phi(w) = \frac{j}{\pi} \ln w$$
$$\Phi(z) = \frac{j}{\pi} \ln \left(j\frac{1-z}{1+z} \right)$$

The flux lines and equipotential lines are found from $\mathfrak{Re}\{\Phi(z)\}$ and $\Im\{\Phi(z)\}$ and are shown in Figure 4.17.

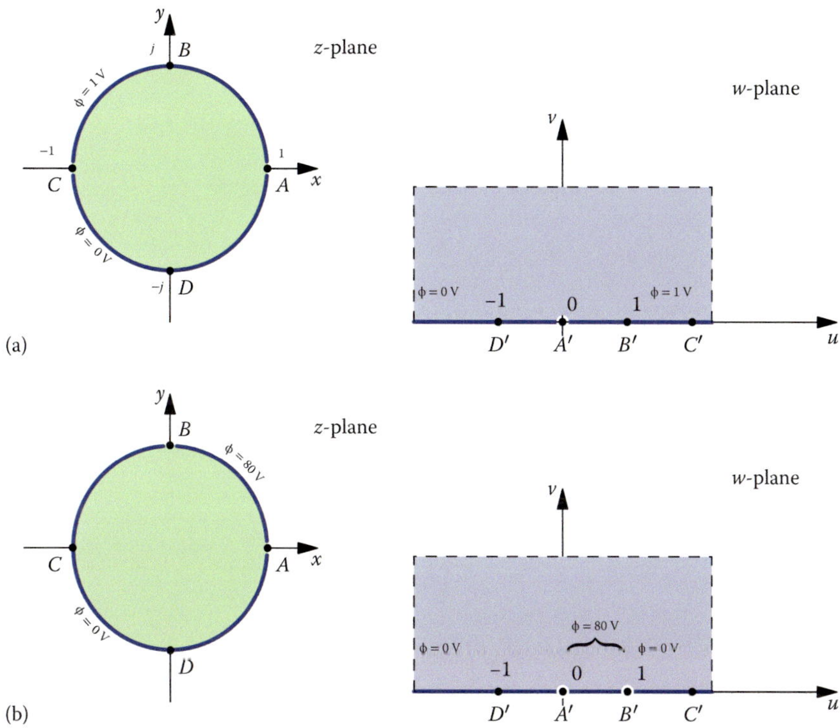

FIGURE 4.16
The bilinear mapping $w = j\frac{1-z}{1+z}$ for the boundary conditions in Example 4.16. (a) The top half is held at 1 V. (b) One quarter of the cylinder is held at 80 V.

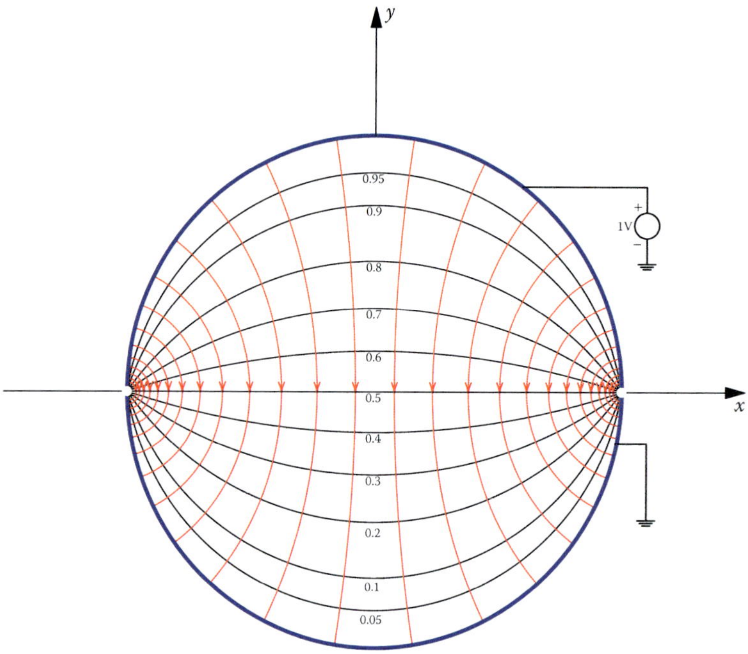

FIGURE 4.17
Equipotential and flux lines in case of two half-cylinders.

(b) Now consider new boundary conditions. After applying the conformal mapping, we find that in the w-plane we have three strips as shown in Figure 4.16b. The complex potential is readily written by inspection as

$$\Phi(w) = \frac{j80}{\pi}\{\ln w - \ln(w-1)\}$$

$$= \frac{j80}{\pi}\left\{\ln\left(\frac{w}{w-1}\right)\right\}$$

$$\Phi(z) = \Phi\left(j\frac{1-z}{1+z}\right)$$

The equipotential lines and the flux lines are found from $\mathfrak{Re}\{\Phi(z)\}$ and $\Im\{\Phi(z)\}$ and are shown in Figure 4.18. □

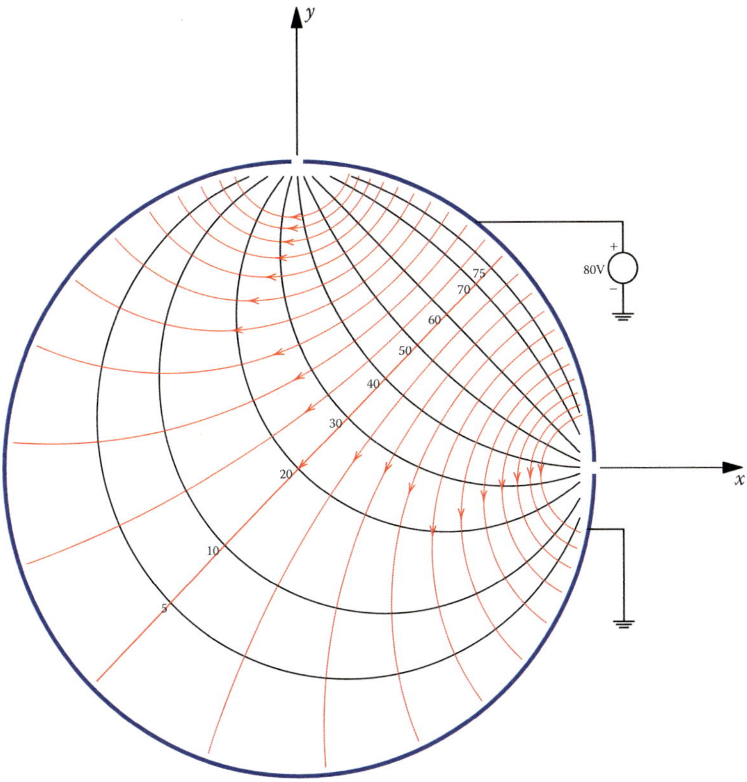

FIGURE 4.18
A quarter of a cylinder at 80 V with respect to the remaining part.

4.7 Evaluation of Capacitance per Unit Length

In the study of transmission lines, electromechanical systems, electrostatic discharge, and so on, knowing the capacitance of a given conductor system is important. If the conductors are uniform in the z-direction, we can treat the problem in two dimensions. The first and the simplest geometry to find capacitance is that of parallel plates assuming no fringing. Other geometries such as coaxial lines, concentric spheres, and so on, could also be done with some algebraic work. But if the shape of conductors does not permit capacitance evaluation, we might attempt the solution using conformal mapping. Whenever we succeed in coming up with a suitable mapping function, the problem could be simplified. But in reality finding a suitable mapping is the hardest part, as we see later. But this method could be successfully employed in certain situations where a mapping function is suggested by intuition or experience. In some difficult problems, this is done in more than one step. Let us see some examples.

Example 4.17. Find an expression for capacitance per unit length in case of a coaxial cable of inner radius a and outer radius b. Let the dielectric medium be homogenous with a dielectric constant ϵ.

Solution: In Example 4.12, we discussed the mapping of logarithmic function and found that a coaxial geometry in the z-plane is mapped into a rectangle in the w-plane. See Figure 4.12. The capacitance per unit length is

$$\begin{aligned}\mathscr{C} &= \epsilon\frac{\psi_1 - \psi_2}{\phi_1 - \phi_2} \\ &= \epsilon\frac{2\pi}{\log b - \log a} \\ &= \frac{2\pi\epsilon}{\log b/a}\end{aligned}$$

□

Example 4.18. Find an expression for the capacitance per unit length in case of two parallel wires each of radius R separated by a distance D. Let the dielectric medium be homogenous with a dielectric constant ϵ.

Solution: We shall take the cross section of the wires in the z-plane. There are two circles each of radius R whose centers are $(\pm D/2, 0)$. If we know an analytic function that maps the two circles into parallel plates, calculation of capacitance is easy. Although this function may not occur to us immediately, we can first map the two circles into two concentric circles using bilinear transformation. Then the logarithmic transformation would finally dispose them into parallel plates. We shall choose an intermediate variable t defined as

$$t = |t|e^{j\text{Arg}t} = \frac{z+l}{z-l}$$

for some $l > 0$. Assuming that $|t|$ is fixed and solving for $|t|^2$, we get

$$|t|^2 = \frac{(x+l)^2 + y^2}{(x-l)^2 + y^2}$$

$$\Rightarrow$$

$$x^2 + y^2 - 2xl\frac{|t|^2+1}{|t|^2-1} + l^2 = 0$$

This equation represents families circles with centers $\left(l\frac{|t|^2+1}{|t|^2-1}, 0\right)$ and radii $\left(\frac{2l|t|}{|t|^2-1}\right)$. The wires under consideration must occupy two of these circles. Let their images in the t-plane be two circles of radii a and b. When $x = 0$, we get $|t| = 1$ meaning that the y-axis is mapped as unit circle. Since the two circles represented by the wires in the z-plane are symmetric to each other with respect to the y-axis, their images should also be symmetric with respect to the unit circle thanks to the property of bilinear mapping. Thus $b = 1/a$. Taking the circle corresponding to the radius $|t| = a$ we can write

$$D/2 = \frac{l(|t|^2 + 1)}{|t|^2 - 1}$$

$$R = \frac{2|t|l}{|t|^2 - 1}$$

from which

$$a = \frac{D}{2R} + \sqrt{\frac{D^2}{4R^2} - 1}$$

Next mapping the t-plane into w-plane using $w = \ln t$, we can convert the coaxial geometry into a parallel plate geometry as was done in Example 4.12. The outer and inner circles in the coaxial problem in the t-plane map to $\ln a$ and $\ln(1/a)$, respectively. We might also recall that

$$\ln a = \cosh^{-1}\left(\frac{D}{2R}\right)$$

These steps are shown in Figure 4.19. The capacitance per unit length is simply given by

$$\begin{aligned} \mathscr{C} &= \epsilon\left(\frac{2\pi}{\ln a - \ln 1/a}\right) \\ &= \frac{\pi\epsilon}{\ln a} \\ &= \frac{\pi\epsilon}{\cosh^{-1}\left(\frac{D}{2R}\right)} \end{aligned}$$

□

Example 4.19. Consider a cable formed by two right circular cylinders of radii a and b one inside the other with their axes parallel and at a distance c apart. (a) Use conformal transformation technique to show that the capacitance per unit length is $\left(\frac{2\pi\epsilon}{\cosh^{-1}\left(\frac{a^2+b^2-c^2}{2ab}\right)}\right)$ (b) Using the numerical values $b = 12, a = 2, c = 4$ and a potential difference of 100 V between inner and the outer cylinders, obtain the complex potential and sketch the equipotential and flux lines in the space between the two cylinders. (c) Show that as $c \to 0$, $\mathscr{C} \to \frac{2\pi\epsilon}{\log b/a}$. (d) Let a and b be fixed but allow c to vary from 0 to $(b-a)$, sketch the variation in $\mathscr{C}$ as a function of c. Mark the line showing the capacitance

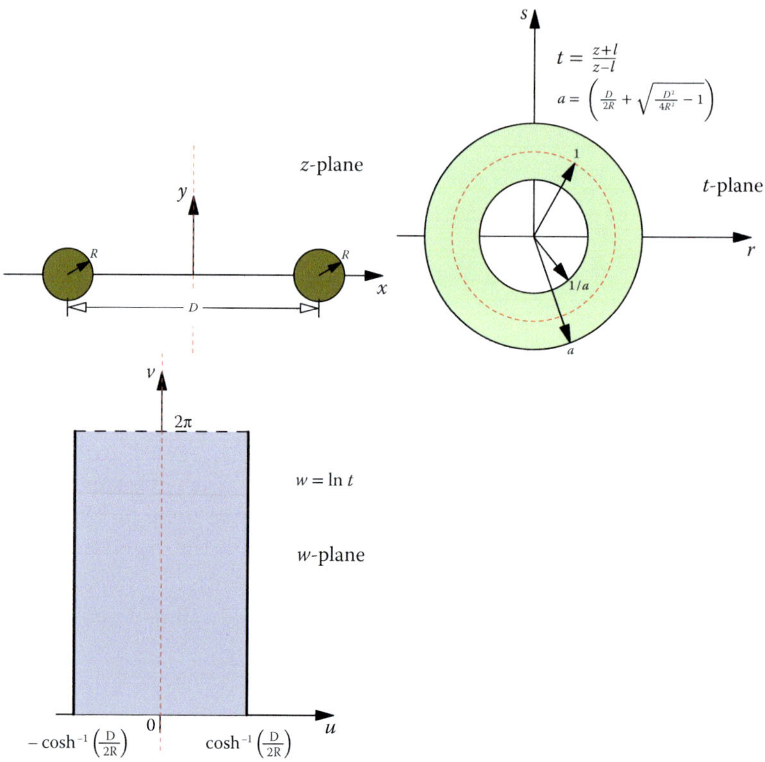

FIGURE 4.19
Transformation of two-wire geometry into parallel plate geometry.

(in pF) of coaxial version and demonstrate that the value of $\mathscr{C}$ approaches $\frac{2\pi\epsilon}{\ln\frac{b}{a}}$. (e) Suppose the inner cylinder is placed outside the outer cylinder, how do you modify the formula of capacitance per unit length? (f) Show that for the case when the inner cylinder is placed outside at symmetrical point and if $a = b \to R$ and $c \to D$, show that $\mathscr{C} \to \frac{\pi\epsilon}{\cosh^{-1}\frac{D}{2R}}$ the expression obtained for two-wire line in Example 4.18.

Solution: (a) The cross-section of the cable is shown in Figure 4.20. We can once again use a bilinear mapping of the form $t = \sigma + j\omega = A\left(\frac{z-z_1}{z-p_1}\right)$ where z_1 and p_1 are the zero and pole or preimages in the z-plane of 0 and ∞ in the t-plane. The constant A is to be chosen suitably. Assume that the inner circle maps to a circle of radius r and the outer circle maps to another concentric circle of radius R. From the property of bilinear mapping, inasmuch as 0 and ∞ are symmetric with respect to every circle around the origin $C(0)$ in the t-plane, even z_1 and p_1 must be symmetric with respect to the preimage

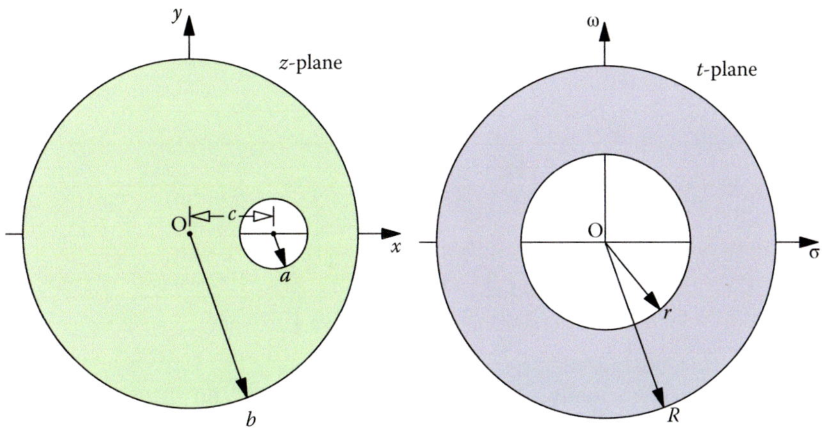

FIGURE 4.20
Transformation of eccentric circles into concentric circles.

thereof. Symmetry with respect to the outer and inner circle in the z-plane requires, respectively,

$$z_1 p_1 = b^2$$
$$(z_1 - c)(p_1 - c) = a^2$$

Since the points $z = c + a$ and $z = b$ map to r and R, respectively,

$$r = A\left(\frac{c + a - z_1}{c + a - p_1}\right) = \frac{Aa}{c - p_1}$$
$$R = A\left(\frac{b - z_1}{b - p_1}\right) = \frac{Az_1}{b}$$

where use has been made of the symmetry of the circles. From the above, we might evaluate the following expression:

$$\frac{1}{2}\left(\frac{R}{r} + \frac{r}{R}\right) = \frac{1}{2}\left(z_1 \frac{p_1 - c}{ab} + \frac{ab}{z_1(p_1 - c)}\right)$$
$$= \frac{a^2 + b^2 - c^2}{2ab}$$

Now that we have mapped the eccentric circles into concentric circles, finding capacitance is obtained from $\mathscr{C} = \frac{2\pi\epsilon}{\ln R/r}$. Recalling that logarithmic and inverse hyperbolic cosine are related by a formula $\ln x = \cosh^{-1} \frac{1}{2}\left(x + \frac{1}{x}\right)$, we can more elegantly write

$$\mathscr{C} = \frac{2\pi\epsilon}{\cosh^{-1}\left(\frac{a^2+b^2-c^2}{2ab}\right)}$$

(b) Since we already developed the expression for the complex potential in case of coaxial cable in Example 4.12, we can readily use it here with appropriate symbols:

$$\Phi(t) = V_o \frac{\ln\left(\frac{t}{R}\right)}{\ln\left(\frac{r}{R}\right)}$$

$$\Phi(z) = V_o \frac{\ln\left(\frac{b(z-z_1)}{z_1}\right)}{\ln\left(\frac{ab}{cz_1-b^2}\right)}$$

where use has been made of the symmetry relations and even z_1 has to be found from them. The potential and the flux functions can be obtained from $\phi(x, y) = \Re\mathfrak{e}\{\Phi(z)\}$ and $\psi(x, y) = \Im\{\Phi(z)\}$, respectively. These are shown in Figure 4.21.

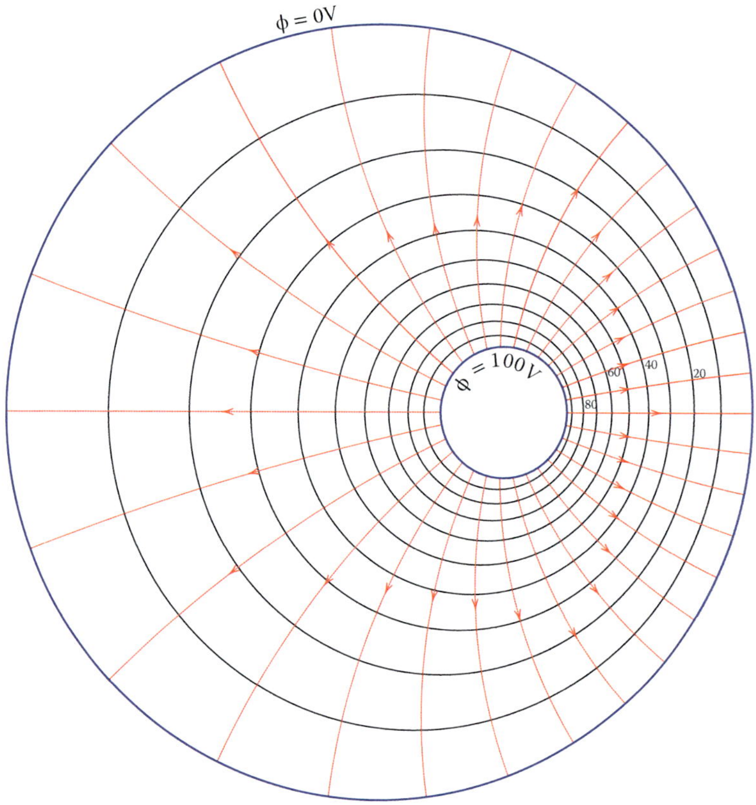

FIGURE 4.21
Equipotential and flux lines in a cable made of eccentric cylinders.

(c) Allowing $c \to 0$, we get

$$\mathscr{C} = \frac{2\pi\epsilon}{\cosh^{-1}\left(\frac{a^2+b^2}{2ab}\right)} = \frac{2\pi\epsilon}{\cosh^{-1}\frac{1}{2}\left(\frac{a}{b}+\frac{b}{a}\right)} = \frac{2\pi\epsilon}{\ln\frac{b}{a}}$$

(d) The capacitance in pF/m is evaluated as function of axial separation c and is plotted in Figure 4.22. Note that as $c \to 0$, the value of the capacitance approaches that of the coaxial cable as expected.

(e) The formula for $\mathscr{C}$ gives meaningful results when the inner cylinder is inside the outer cylinder and $c < (b-a)$ or (squaring and rearranging) for $(a^2+b^2-c^2) > 2ab$. Suppose the inner cylinder is taken out and placed at a symmetric distance, $c = D$, outside the outer cylinder. This inequality would reverse. Hence the expression for the capacitance per unit length is given by

$$\mathscr{C} = \frac{2\pi\epsilon}{\cosh^{-1}\left(\frac{c^2-a^2-b^2}{2ab}\right)}$$

(f) Substituting $a = b = R,\ c = D$ the capacitance per unit length would become,

$$\mathscr{C} = \frac{2\pi\epsilon}{\cosh^{-1}\left(\frac{D^2-2R^2}{2R^2}\right)} = \frac{2\pi\epsilon}{2\cosh^{-1}\left(\frac{D}{2R}\right)} = \frac{\pi\epsilon}{\cosh^{-1}\left(\frac{D}{2R}\right)}$$

which is the same as the expression for capacitance per unit length of a two-wire line. □

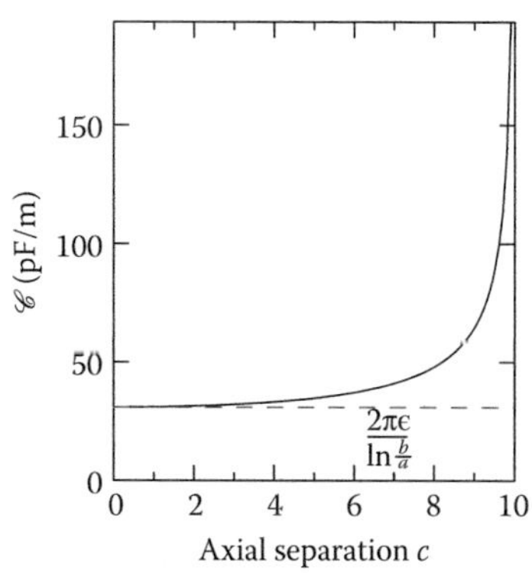

FIGURE 4.22
Capacitance per meter as a function of axial separation c. Note that as c decreases to zero it reaches the value of a coaxial cable.

Example 4.20. (a) A transmission line is formed by two confocal elliptical cylinders with foci at $(\pm c, 0)$, their semimajor axes being a and b, $(b > a)$. Develop an expression for $\mathscr{C}$. (b) Allow $c \to 0$ and show that $\mathscr{C} \to \frac{2\pi\epsilon}{\ln\frac{b}{a}}$. (c) Suppose the inner elliptical conductor reduces to a strip with edges at the foci, find an expression for $\mathscr{C}$.

Solution: (a) Let us consider the analytic function $w = \cosh^{-1}\frac{z}{c}$. By substituting $z = x + jy$ and $w = u + jv$, we can obtain the following relations:

$$\begin{aligned} z &= c\cosh w \\ x + jy &= c\cosh(u + jv) \\ &\Rightarrow \\ x &= c\cosh u \cos v \\ y &= c\sinh u \sin v \end{aligned}$$

Eliminating v, we find that

$$\left(\frac{x}{c\cosh u}\right)^2 + \left(\frac{y}{c\sinh u}\right)^2 = 1$$

which represents family of ellipses in the z-plane with semimajor axis of $c\cosh u$. Thus, the two given ellipses are mapped into parallel lines $u = \cosh^{-1}\frac{a}{c}$ and $u = \cosh^{-1}\frac{b}{c}$ as depicted in Figure 4.43. This permits us to express $\mathscr{C}$ readily as

$$\begin{aligned} \mathscr{C} &= \frac{2\pi\epsilon}{\cosh^{-1}\frac{b}{c} - \cosh^{-1}\frac{a}{c}} \\ &= \frac{2\pi\epsilon}{\ln\left(\frac{\frac{b}{c}+\sqrt{\frac{b^2}{c^2}-1}}{\frac{a}{c}+\sqrt{\frac{a^2}{c^2}-1}}\right)} \\ &= \frac{2\pi\epsilon}{\ln\left(\frac{b+\sqrt{b^2-c^2}}{a+\sqrt{a^2-c^2}}\right)} \end{aligned}$$

(b) Allowing $c \to 0$, we find that $\mathscr{C} \to \frac{2\pi\epsilon}{\ln\frac{b}{a}}$ and reduces to that of a coaxial cable as we expect.

(c) If a strip is placed inside an elliptical conductor, we can modify $\mathscr{C}$ simply by setting $c = a$. Thus,

$$\begin{aligned} \mathscr{C} &= \frac{2\pi\epsilon}{\ln\left(\frac{b+\sqrt{b^2-a^2}}{a}\right)} \\ &= \frac{2\pi\epsilon}{\cosh^{-1}\frac{b}{a}} \end{aligned}$$

□

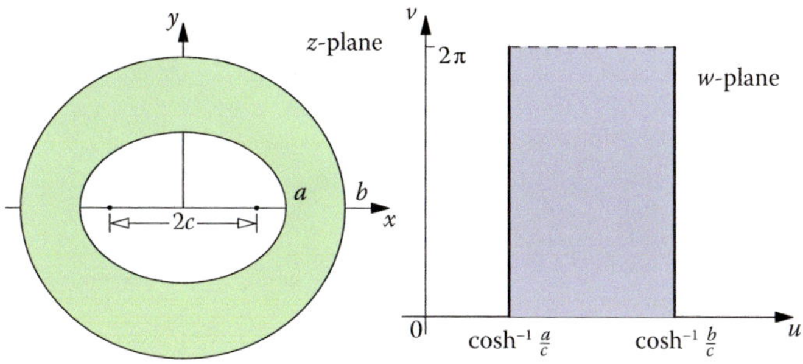

FIGURE 4.23
Mapping an elliptical geometry into a parallel plate geometry using $w = \cosh^{-1}\frac{z}{c}$.

4.8 The Schwarz–Christoffel Transformation

The Schwarz–Christoffel transformation (SCT) is a conformal transformation $w = f(z)$ that maps the real axis into a general polygon from z-plane to w-plane. This tool is extensively used in finding the capacitance per unit length of transmission lines of several shapes. First, we shall formally derive this transformation and then show how it can be used through several examples. Consider a curve $C : z(t) = x(t) + jy(t)$ in the z-plane and let its image be $K : w(t) = u(t) + jv(t)$ in the w-plane. A unit tangent to C at t_0 is

$$\boldsymbol{\tau} = z'(t_0) = x'(t_0) + jy'(t_0) \tag{4.30}$$

and the corresponding tangent in the w-plane at $w_0 = f(z_0)$ is

$$\mathbf{T} = w'(t_0) = u'(t_0) + jv'(t_0) \tag{4.31}$$

If the angle of inclination of $\boldsymbol{\tau}$ at t_0 is β, the corresponding angle of $\mathbf{T}$ is given by

$$\arg \mathbf{T} = \arg f'(z_0)z'(t_0) = \arg f'(z_0) + \beta \tag{4.32}$$

As shown in Figure 4.24 this implies that the tangent in the w-plane is turned by an angle $\arg f'(z)$ for each point z in the z-plane. Now suppose

$$f'(z) = A(z - x_1)^{-k_1}(z - x_1)^{-k_1} \cdots (z - x_1)^{-k_N} = A\prod_{i=1}^{N}(z - x_i)^{-k_i} \tag{4.33}$$

where A is an arbitrary constant, the set of points $\{x_1, x_2 \cdots x_N\}$ a partition on the real line along which $\arg \boldsymbol{\tau} = 0$ everywhere, and each k_i is a real number. The angle of $\mathbf{T}$ becomes

$$\arg \mathbf{T} = \arg f'(z) = \arg A - \sum_{i=1}^{N} k_i \arg (z - x_i) \tag{4.34}$$

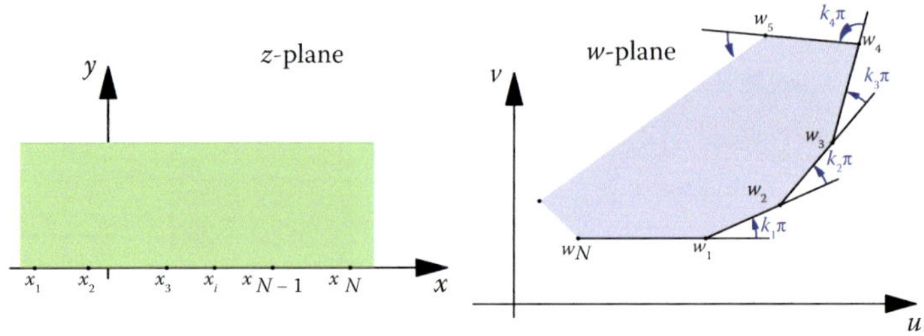

FIGURE 4.24
The Schwarz–Christoffel transformation.

As z varies along the real line, we note that

$$\arg\,(z - x_i) = \begin{cases} \pi & \text{for } x < x_i \\ 0 & \text{for } x > x_i \end{cases} \tag{4.35}$$

and with each crossing of x_i, the angle of $\mathbf{T}$ undergoes a step change of $k_i\pi$. Thus, under an SCT, $f(z)$ eventually traces a polygon P such as the one shown in Figure 4.24 with the exterior angle $k_i\pi$ at w_i. The image of the partition $\{x_1, x_2 \cdots x_N\}$ happens to be the vertices of the polygon $\{w_1, w_2, \cdots w_N\}$, and the upper half of the z-plane maps onto the interior of P. Note further that the upper half of the z-plane as well as the interior of the polygon happen to be on the left side of an observer walking along the real line of the z-plane or on the polygon in the w-plane. Since the exterior angles of a closed polygon sum up to 2π, we find that

$$\sum_{i=1}^{N} k_i = 2 \tag{4.36}$$

Integrating $f'(z)$ along the real line, we can write the general form of an SCT as

$$w = B + A \int^{z} \prod_{i=1}^{N} (z - x_i)^{-k_i} \mathrm{d}z \tag{4.37}$$

where B is another constant that determines the location of the polygon in the w-plane, whereas, the constant A affects the size and orientation thereof. Also, the lower limit of integration is inconsequential and is left out. To apply SCT for the transmission line problems, let us take few examples.

Example 4.21. Consider a pair of inclined half-planes making an angle of α held at a potential difference of V_0. Assume that $\alpha \neq m\pi$, for $m = 0, 1, 2 \ldots$ and that an infinitesimal gap exists between the plates at w_o where they seem to meet. Use SCT and obtain an expression for the potential at any general point between the plates.

Solution: The given geometry is trivial if the gap is at the origin. However, we shall pretend that we do not know the solution and try to get it using SCT. We might treat the geometry as a polygon with one vertex at w_o and let us map the geometry into two coplanar half-planes on either side of the origin as shown in Figure 4.25. The angle between the plates is α, so the external angle is $\pi - \alpha$. Hence using SCT,

$$\begin{aligned} w &= C_2 + C_1 \int^z z^{-\frac{(\pi-\alpha)}{\pi}} \mathrm{d}z \\ &= C_2 + C_1 \frac{\pi}{\alpha} z^{\frac{\alpha}{\pi}} \end{aligned}$$

The constraint $z = 0$ for $w = w_o$ yields $C_2 = w_o$. So the required mapping is

$$\begin{aligned} w &= w_o + C_1 \frac{\pi}{\alpha} z^{\frac{\alpha}{\pi}} \\ w - w_o &= C_1 \frac{\pi}{\alpha} z^{\frac{\alpha}{\pi}} \\ \mathrm{Arg}(w - w_o) &= \frac{\alpha}{\pi} \mathrm{Arg}(z) \\ \mathrm{Arg}(z) &= \frac{\pi}{\alpha} \mathrm{Arg}(w - w_o) \end{aligned}$$

Using the results of Example 4.13 we know that the expression for potential is $V_0/\pi \mathrm{Arg}(z)$. Taking argument on either side assuming C_1 to be a real arbitrary constant,

$$\begin{aligned} \phi(x, y) &= \frac{V_0}{\pi} \mathrm{Arg} z \\ \phi(u, v) &= \frac{V_0}{\alpha} \mathrm{Arg}(w - w_o) \end{aligned}$$

□

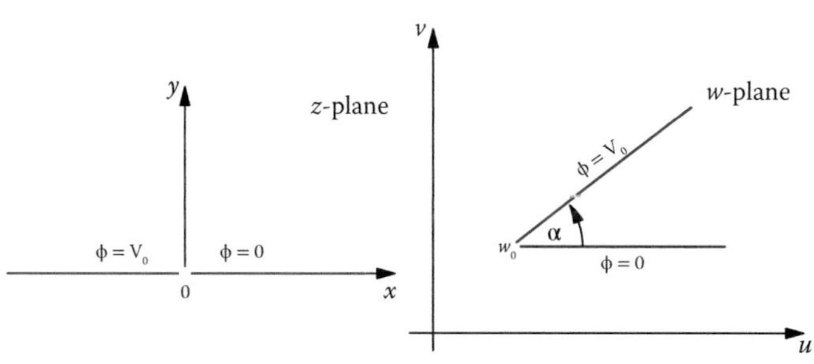

FIGURE 4.25
Potential between inclined half-planes.

Example 4.22. An infinite trough is made of a strip of width a lying in the xz-plane and the semi-infinite strips $x = 0,\ y > 0$ and $x = a,\ y > 0$ with an infinitesimal gap at the corners $(0, 0, 0$ and $(a, 0, 0)$. The strip $0 < x < a,\ y = 0$ is held at a potential V_0 while the side walls are grounded. Find an expression for the potential in the region between the plates. Also develop a solution using series method and compare your answer.

Solution: Let us transform the given trough from the z-plane into a strip between two coplanar planes in the t-plane ($t = r+js$) as shown in Figure 4.26. We recognize that there are two right angled corners in the z-plane at $x = 0$ and $x = a$, which we might send to $t = \pm 1$. Using SCT,

$$\begin{aligned} z &= C_2 + C_1 \int^t (t+1)^{-\frac{1}{2}}(t+1)^{-\frac{1}{2}}\,\mathrm{d}t \\ &= C_2 + C_1 \int^t \frac{\mathrm{d}t}{\sqrt{t^2-1}} \\ &= C_2 + C_1 \cosh^{-1} t \end{aligned}$$

Using the conditions that $z = 0,\ a$ go to $t = \pm 1$, we can rewrite the above as

$$z = a + \frac{ja}{\pi}\cosh^{-1} t$$

Expressing t in terms of z, we get

$$t = 1 - 2\cos^2\frac{\pi z}{2a}$$

We can view the boundary value problem in the t-plane as a case of Dirichlet's problem discussed in Example 4.15 and we can readily write the solution for

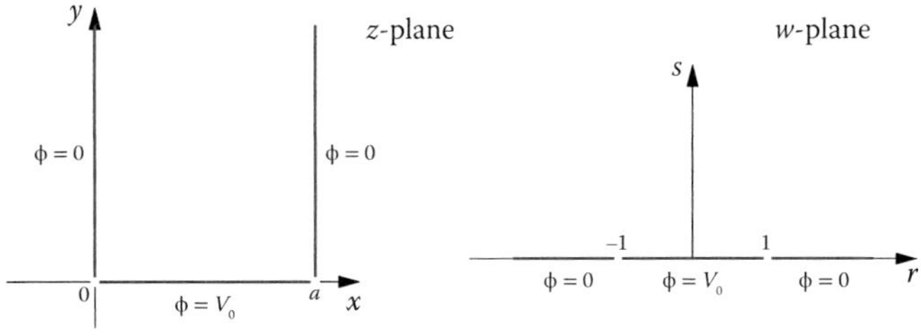

FIGURE 4.26
An infinite trough.

potential as

$$\begin{aligned}
\phi(r,s) &= -\frac{V_0}{\pi}\mathrm{Arg}(t+1) + \frac{V_0}{\pi}\mathrm{Arg}(t-1)\\
&= \frac{V_0}{\pi}\mathrm{Arg}\left(\frac{t-1}{t+1}\right)\\
\phi(x,y) &= \frac{V_0}{\pi}\mathrm{Arg}\left(-\cot^2\frac{\pi z}{2a}\right)\\
&= \frac{2V_0}{\pi}\mathrm{Arg}\left(j\cot\frac{\pi z}{2a}\right)\\
&= \frac{2V_0}{\pi}\mathrm{Arg}\left(\sinh\frac{\pi y}{a} + j\sin\frac{\pi x}{a}\right)\\
&= \frac{2V_0}{\pi}\tan^{-1}\left(\frac{\sin\frac{\pi x}{a}}{\sinh\frac{\pi y}{a}}\right)
\end{aligned}$$

The complex potential can be written as

$$\Phi(z) = \frac{-2jV_0}{\pi}\ln\left(j\cot\frac{\pi z}{2a}\right) \tag{4.38}$$

Solution via Fourier series: The solution we obtained through conformal mapping can also be attempted using the Fourier series technique. From the nature of the boundary conditions, a basic solution that remains zero for $x = 0$ and $x = a$ must be of the form $\sin k_x x$ and further, if it has to decay to zero for $y = \infty$ from an initial value of V_0 at $y = 0$, it should contain a factor of the form $\mathrm{e}^{-k_y y}$. So we start out with a fundamental solution of the form $\mathrm{e}^{-k_y y}\sin k_x x$. In order that $\phi(x,y)$ satisfy the Laplace equation in two dimensions, we require that $\nabla^2\phi = 0$. This necessitates that $k_x = k_y$, and the boundary conditions at $x = 0$ and $x = a$ require that $k_x = n\pi/a$ for integer values of n. Thus, a general series expression for the potential can be built using

$$\phi(x,y) = \sum_{n=1}^{\infty} C_n \mathrm{e}^{-\frac{n\pi y}{a}}\sin\frac{n\pi x}{a}$$

Finally, to satisfy the constraint that $\phi(x,0) = V_0$ we need to write,

$$\phi(x,0) = V_0 = \sum_{n=1}^{\infty} C_n \sin\frac{n\pi x}{a}$$

From the knowledge of Fourier series, we can readily obtain an expression for C_n as

$$C_n = \begin{cases}\frac{4V_0}{\pi n} & \text{for} \quad n \quad \text{odd}\\ 0 & \text{for} \quad n \quad \text{even}\end{cases}$$

which permits us to write

$$\phi(x, y) = \frac{4V_0}{\pi} \sum_{n \text{ odd}} \frac{1}{n} e^{-\frac{n\pi y}{a}} \sin \frac{n\pi x}{a}$$

This can be viewed as the imaginary part of a complex function of $z = x + jy$ as follows:

$$\phi(x, y) = -\frac{4V_0}{\pi} \mathfrak{Re} \left[\sum_{n \text{ odd}} \frac{j}{n} e^{\frac{jn\pi z}{a}} \right]$$

Now let us define $\zeta = e^{\frac{j\pi z}{a}}$. We can write

$$\phi(x, y) = -\frac{4V_0}{\pi} \mathfrak{Re} \left[\sum_{n \text{ odd}} j \frac{\zeta^n}{n} \right]$$

Here for all $y > 0$, we note that $|\zeta| = e^{-\pi y/a} < 1$. So to simplify our result further, we can recall the series

$$\begin{aligned}
\ln(1 + \zeta) &= \zeta - \frac{1}{2}\zeta^2 + \frac{1}{3}\zeta^3 - \frac{1}{4}\zeta^4 + \ldots \\
\ln(1 - \zeta) &= -\zeta - \frac{1}{2}\zeta^2 - \frac{1}{3}\zeta^3 - \frac{1}{4}\zeta^4 - \ldots \\
\frac{1}{2} \ln \left(\frac{1 + \zeta}{1 - \zeta} \right) &= \zeta + \frac{1}{3}\zeta^3 + \frac{1}{5}\zeta^5 + \ldots
\end{aligned}$$

valid for $|\zeta| < 1$. So we can write the potential function as

$$\phi(x, y) = -\frac{2V_0}{\pi} \mathfrak{Re} \left[j \ln \left(\frac{1 + \zeta}{1 - \zeta} \right) \right]$$

Substituting back $\zeta = e^{j\pi z/a}$, we get

$$\begin{aligned}
\phi(x, y) &= \mathfrak{Re} \left[-\frac{2jV_0}{\pi} \ln \left(j \cot \frac{\pi z}{a} \right) \right] \\
&= \mathfrak{Re} \{\Phi(z)\}
\end{aligned}$$

where $\Phi(z)$ tallies with Equation 4.38. The equipotential and the flux lines for the infinite trough are shown in Figure 4.27 for $V_0 = 10$ V. □

Example 4.23. Two coplanar strips of finite width described by $z = 0,\ a < |x| < b$ are held at $+V_0$ and $-V_0$ as shown in Figure 4.28. Use SCT and obtain a sketch of equipotential and flux lines for the case $a = 1, b = 5, V_0 = 10$.

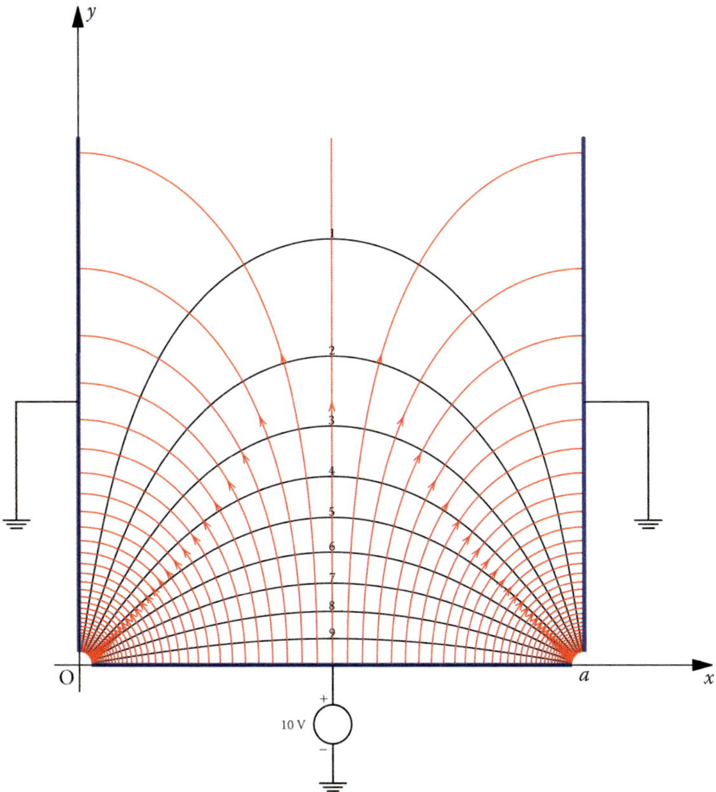

FIGURE 4.27
Equipotential and flux lines in the infinite trough of Example 4.22.

Solution: Let us transform the geometry of the given strips in the z-plane into parallel planes in the w-plane into using SCT. The transformation shown in Figure 4.28 is given by

$$w = C_2 + C_1 \int^z \frac{\mathrm{d}z}{(z^2 - a^2)(z^2 - b^2)}$$

In dealing with integrals of the above type, it is apt to introduce the incomplete Jacobi elliptic function defined in the following way for finite ground plane. The reader is encouraged to consult [3], [6], and [8] for additional details.

$$u = \int_0^x \frac{\mathrm{d}t}{\sqrt{1-t^2}\sqrt{1-k^2t^2}} \tag{4.39a}$$

$$x = \mathrm{sn}(u) \tag{4.39b}$$

$$u = \mathrm{sn}^{-1}(x) \tag{4.39c}$$

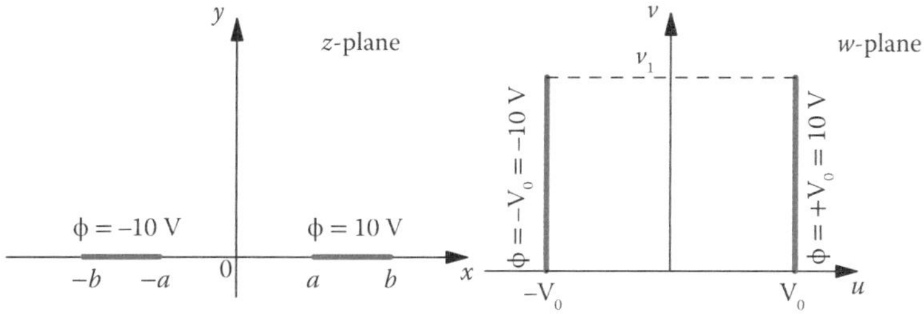

FIGURE 4.28
Two coplanar strips of *finite* width being mapped to a parallel plate geometry.

For a more elaborate discussion of these special functions and their properties, we encourage the reader to consult [7]. Using this notation we can further write,

$$w = C_2 + \frac{C_1}{a}\text{sn}^{-1}\left(\frac{z}{a}, \frac{a}{b}\right)$$

Requiring that the points $(0 + j0), (\pm b + j0), (\pm a + j0)$ be mapped to $(0 + j0), (\pm u_0 + jv_1), (\pm V_0 + j0)$ we can rewrite the mapping functions as

$$\begin{aligned} w &= V_0 \frac{\text{sn}^{-1}\left(\frac{z}{a}, \frac{a}{b}\right)}{\text{sn}^{-1}\left(1, \frac{a}{b}\right)} \\ &= V_0 \frac{\text{F}\left(\sin^{-1}\frac{z}{a}, \frac{a}{b}\right)}{\text{K}\left(\frac{a}{b}\right)} \end{aligned}$$

Although the above discussion holds for the upper half-plane of the z-plane, by analytic continuation and the existing symmetry, we might take a copy of the picture of fields in the upper half-plane and paste in the lower half-plane. The complete picture is shown in Figure 4.29.

The capacitance per unit length of this structure is easily found from the w-plane as

$$\begin{aligned} \mathscr{C} &= \frac{\epsilon v_1}{2V_0} \\ &= \Im\left\{\frac{\epsilon\text{F}\left(\sin^{-1}\left(\frac{b}{a}, \frac{a}{b}\right)\right)}{2\text{K}(a/b)}\right\} \end{aligned}$$

□

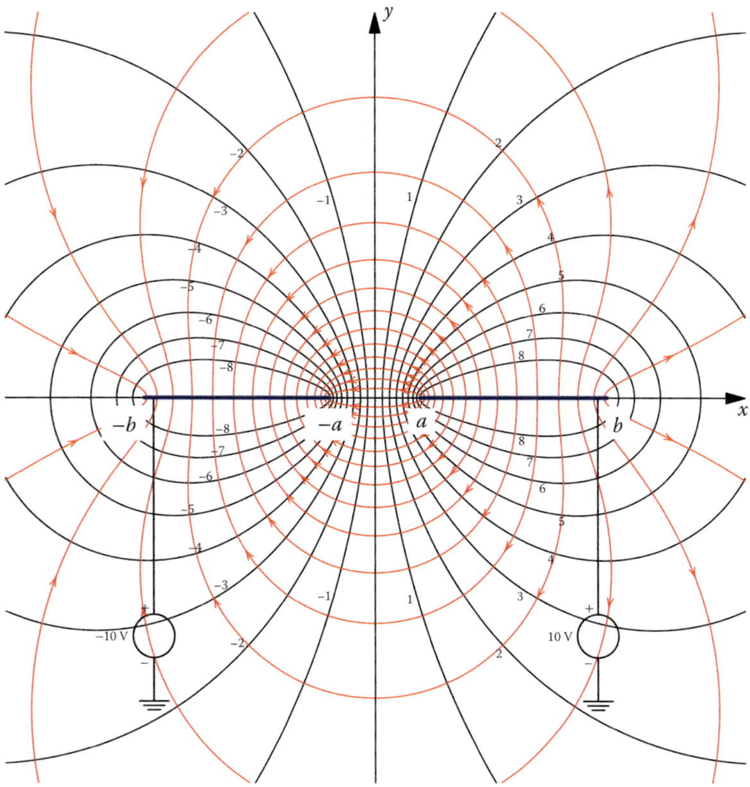

FIGURE 4.29
Equipotential and flux lines of a pair of coplanar strips of finite width.

4.9 Strip Lines and Microstrip Lines

Some of the developments that took place in the second half of the twentieth century were phenomenal. Semiconductor device fabrication, surface mount technology, and microminiaturization of electronic circuits ushered in a new era of small size, light weight electronic devices. This was also coupled with the invention of printed circuit board technology that brought in planar transmission lines to supplant bulky waveguide structures. Strip transmission lines were introduced in the early 1950s. They consist of a flat thin strip conductor sandwiched between two dielectric sheets with conductive plating on both the outer sides. A simplified variant of a strip line is the microstrip line that consists of a dielectric with a strip on one side and a ground plane on the other side. Contrary to the strip line, a microstrip line is an open structure needing a high dielectric constant substrate to have a tighter coupling of fields closer to the strip conductor.

In this section, we analyze a strip line using the conformal transformation technique. First, we consider a strip line with infinite ground plane and then modify the same for finite ground plane. After this, we consider a more general case of having an elliptical strip between two ground planes. This permits us to obtain formulas for a strip line with a horizontal strip as well as a vertical strip. The latter may not be of practical importance, but it helps us to appreciate the results from a mathematical point of view.

4.9.1 Strip Line

A strip transmission line with the strip placed symmetrically between two parallel conductors is shown in Figure 4.30. In a more general case, the strip may not be placed symmetrically between the ground planes.

First, we shall assume that the width of each ground plane is infinite. If the width of the strip is small compared with that of the ground planes, this assumption is justified. In the next section, however, we discuss the effect of ground plane width more accurately. A moment's observation reveals that the symmetry in the strip line structure permits us to analyze only one-fourth of its cross section. As shown in Figure 4.31, it is taken from the w-plane to the upper half of the z-plane using SCT.

The mapping of various points is shown in Table 4.1.

The mapping function is given by

$$w = B + A\int^{z}(z-x_1)^{-1/2}(z-x_2)^{-1/2}(z-x_3)^{-1/2}(z-x_4)^{-1/2}\mathrm{d}z \quad (4.40)$$

$$= B + \frac{A}{\sqrt{x_1x_4}}\int^{z}\left[\left(1-\frac{z}{x_1}\right)\left(1-\frac{z}{x_4}\right)(z-x_2)(z-x_3)\right]^{-1/2}\mathrm{d}z \quad (4.41)$$

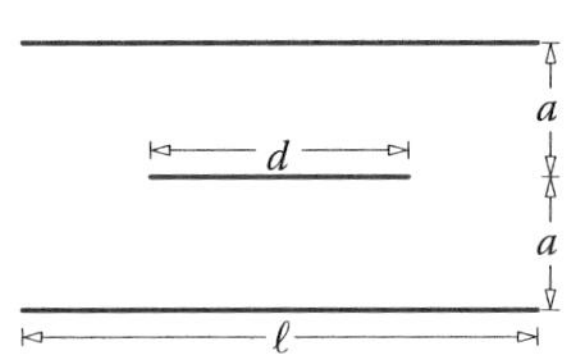

FIGURE 4.30
Strip line geometry.

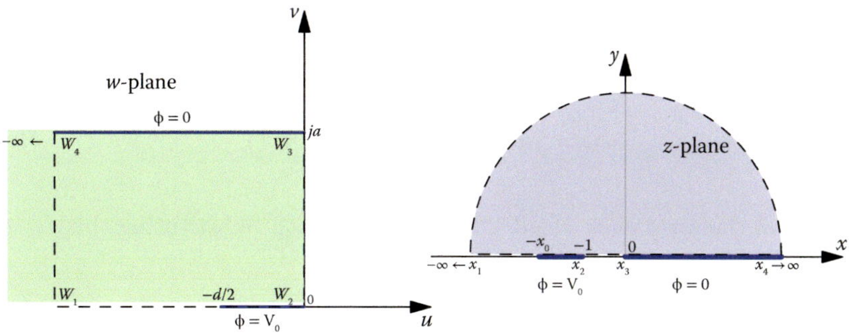

FIGURE 4.31
Mapping between w- and z-planes.

TABLE 4.1
Showing the Mapping of Various Points in Figure 4.31

Point in the w-plane	Point in the z-plane	Angle k_i/π
$W_1 = -\infty$	$x_1 = -\infty$	$\pi/2$
$W_2 = 0$	$x_2 = -1$	$\pi/2$
$W_3 = ja$	$x_3 = 0$	$\pi/2$
$W_4 = -\infty + ja$	$x_4 = \infty$	$\pi/2$
$-d/2$	$-x_0$	match point

Let $x_1 \to \infty$, $x_4 \to \infty$, but $\frac{A}{\sqrt{x_1 x_4}} \to A'$. Using the values of x_2 and x_3, we can write the above as

$$\begin{aligned} w &= B + A' \int^z [(z - x_2)(z - x_3)]^{-1/2}\, \mathrm{d}z \\ &= B + A' \int^z \frac{1}{\sqrt{z(z+1)}}\, \mathrm{d}z \\ &= B + 2A' \ln\left(\sqrt{z} + \sqrt{z+1}\right) \end{aligned} \tag{4.42}$$

Substituting $w = w_2 = 0, z = x_2 = -1$ and $w = w_3 = ja, z = x_3 = 0$, we can determine the constants A' and B and express the mapping as

$$w = ja - \frac{2a}{\pi} \ln\left(\sqrt{z} + \sqrt{z+1}\right) \tag{4.43}$$

The tip of the strip at $w = -d/2$ maps to $z = -x_0$. Substituting this

$$\frac{-d}{2} = ja - \frac{2a}{\pi} \ln\left(\sqrt{x_0} + \sqrt{x_0+1}\right) \tag{4.44}$$

Rearranging terms and solving for x_0, we get

$$x_0 = \cosh^2 \frac{\pi d}{4a} \tag{4.45}$$

We have just mapped the strip geometry into that of two coplanar conductors. We still do not know its capacitance. So we attempt another mapping to change the coplanar conductors into parallel plate geometry. This is shown in Figure 4.32. The plate separation is v_0 and the width is unity. So the capacitance per unit length becomes ϵ_0/v_0. The mapping between various points in the w'-plane and the z-plane is shown in Table 4.2. An analytic function that carries out the above mapping is of the form

$$w' = B_0 + A_1 \int^z \left[(\xi + x_0)(\xi + 1)(\xi)\right]^{-1/2} \mathrm{d}\xi \tag{4.46}$$

$$= B_0 + A_1 \int^z \frac{\mathrm{d}\xi}{\sqrt{(\xi + x_0)(\xi+1)(\xi)}} \tag{4.47}$$

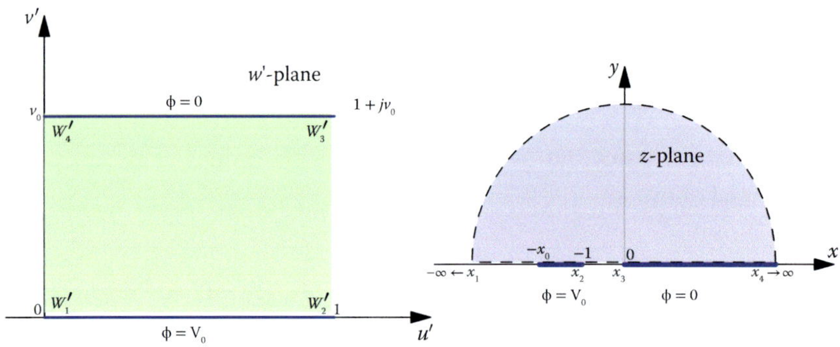

FIGURE 4.32
Mapping between w'- and z-planes.

TABLE 4.2
Mapping between Various Points in the w'-Plane and the z-Plane

Point in the w'-plane	Point in the z-plane	Angle k_i/π
$W_1' = 0$	$x_1 = -x_0$	$\pi/2$
$W_2' = 0$	$x_2 = -1$	$\pi/2$
$W_3' = 1 + v_0$	$x_3 = 0$	$\pi/2$
$W_4' = jv_0$	$x_4 = \infty$	$\pi/2$
$-d/2$	$-x_0$	Match point

Let us make the substitution $\xi = -\lambda^2$ and $d\xi = -2\lambda d\lambda$.

$$w' = B_0 + A_1 \int^{\lambda=\sqrt{-z}} \frac{-2\lambda d\lambda}{\sqrt{-\lambda^2(1-\lambda^2)(x_0-\lambda^2)}} \tag{4.48}$$

$$= B_0 + \frac{2A_1 j}{\sqrt{x_0}} \int^{\lambda=\sqrt{-z}} \frac{d\lambda}{\sqrt{(1-\lambda^2)(1-\lambda^2/x_0)}} \tag{4.49}$$

$$= B_0 + \frac{2A_1 j}{\sqrt{x_0}} \mathrm{sn}^{-1}\left(\sqrt{-z}, \frac{1}{\sqrt{x_0}}\right) \tag{4.50}$$

$$= B_0 + A_0 \mathrm{sn}^{-1}\left(\sqrt{-z}, \frac{1}{\sqrt{x_0}}\right) \tag{4.51}$$

where $A_0 = \frac{2A_1 j}{\sqrt{x_0}}$. Recall the definition of the function $\mathrm{sn}(\cdot)$ and $\mathrm{sn}^{-1}(\cdot)$ from 4.39. Substituting the points from Table 4.2 we get

$$0 = B_0 + A_0 \mathrm{sn}^{-1}\left(\sqrt{x_0}, \frac{1}{\sqrt{x_0}}\right) \tag{4.52}$$

$$1 = B_0 + A_0 \mathrm{sn}^{-1}\left(1, \frac{1}{\sqrt{x_0}}\right) \tag{4.53}$$

$$1 + jv_0 = B_0 \tag{4.54}$$

Solving the above equations, we get

$$v_0 = \frac{-j\ \mathrm{sn}^{-1}\left(1, \frac{1}{\sqrt{x_0}}\right)}{\mathrm{sn}^{-1}\left(\sqrt{x_0}, \frac{1}{\sqrt{x_0}}\right) - \mathrm{sn}^{-1}\left(1, \frac{1}{\sqrt{x_0}}\right)} \tag{4.55}$$

From the identities of complete elliptic functions

$$\mathrm{sn}^{-1}(1, k) = K(k) \tag{4.56}$$

$$\mathrm{sn}^{-1}(1/k, k) = K(k) - jK'(k) \tag{4.57}$$

$$K'(k) = K(k') \tag{4.58}$$

$$k' = \sqrt{1 - k^2} \tag{4.59}$$

$$v_0 = \frac{-j\,K\left(\frac{1}{\sqrt{x_0}}\right)}{K\left(\frac{1}{\sqrt{x_0}}\right) - jK'\left(\frac{1}{\sqrt{x_0}}\right) - K\left(\frac{1}{\sqrt{x_0}}\right)} \tag{4.60}$$

$$= \frac{K\left(\frac{1}{\sqrt{x_0}}\right)}{K'\left(\frac{1}{\sqrt{x_0}}\right)} \tag{4.61}$$

$$= \frac{K\left(\mathrm{sech}\frac{\pi d}{4a}\right)}{K\left(\tanh\frac{\pi d}{4a}\right)} \tag{4.62}$$

Thus, the capacitance per unit length is given by

$$\mathscr{C} = \frac{4\epsilon_0 K\left(\tanh\frac{\pi d}{4a}\right)}{K\left(\mathrm{sech}\frac{\pi d}{4a}\right)} \tag{4.63}$$

4.10 Strip with Finite Ground Plane

Let us repeat the above exercise assuming the width of the ground plane to be finite. Again we shall show one-fourth of the geometry in the w-plane and its mapping in the z-plane in Figure 4.33. Using SCT, we can write the transformation as

$$w = C_2 + C_1 \int^z (z - B)^{-1/2}(z - C)^{-1/2}(z - D)(z - E)^{-1/2}\mathrm{d}z \tag{4.64}$$

$$= C_2 + C_1 \int^z \frac{\left(z - k_1^{-2}\right)\mathrm{d}z}{\sqrt{z(z-1)(z-k^{-2})}} \tag{4.65}$$

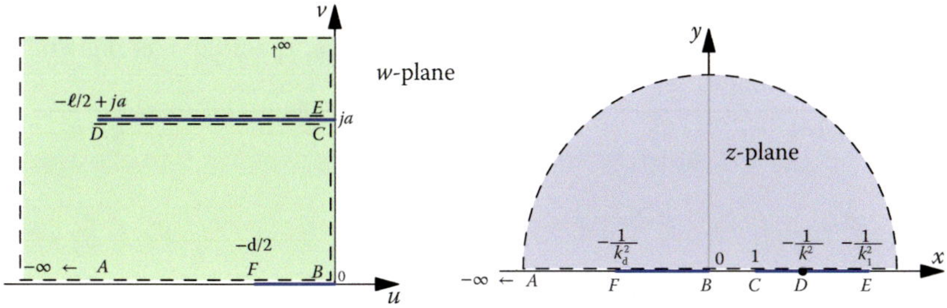

FIGURE 4.33
The case of finite width ground plane. Mapping between w- and z-planes.

Let us put $z = \sin^2 \beta, \; \mathrm{d}z = 2\sqrt{z(1-z)}\mathrm{d}\beta$, we get

$$w = C_2 + \frac{2C_1}{k} \int^{\beta} \frac{\left(k^2 \sin^2 \beta - k^2/k_1^2\right) \mathrm{d}\beta}{\sqrt{1 - k^2 \sin^2 \beta}} \tag{4.66}$$

$$= C_2 - \frac{2C_1}{k} \int^{\beta} \frac{\left(1 - k^2 \sin^2 \beta\right) + \left(k^2/k_1^2 - 1\right) \mathrm{d}\beta}{\sqrt{1 - k^2 \sin^2 \beta}} \tag{4.67}$$

$$= C_2 - \frac{2C_1}{k} \int^{\beta} \left(\left(1 - k^2 \sin^2 \beta\right) + \frac{\left(k^2/k_1^2 - 1\right)}{\sqrt{1 - k^2 \sin^2 \beta}} \right) \mathrm{d}\beta \tag{4.68}$$

$$= C_2 + \frac{2C_1}{k} \left[\left(1 - k^2/k_1^2\right) F(\beta, k) - E(\beta, k)\right] \tag{4.69}$$

The above is the transformation that takes the strip line geometry to the z-plane. The various points in the w-plane, z-plane, corresponding β values and the angles of the polygon are shown in Table 4.3.

We recognize five unknowns *viz.*, C_1, C_2, k, k_1, and k_d and we need five equations to solve for them. Substitution of the values in Table 4.3 generates the following.

$$0 = C_2 + \frac{2C_1}{k} \left[\left(1 - k^2/k_1^2\right) F(0, k) - E(0, k)\right] \tag{4.70}$$

$$0 + ja = C_2 + \frac{2C_1}{k} \left[\left(1 - k^2/k_1^2\right) F(\pi/2, k) - E(\pi/2, k)\right] \tag{4.71}$$

$$-\frac{l}{2} + ja = C_2 + \frac{2C_1}{k} \left[\left(1 - k^2/k_1^2\right) F(\sin^{-1} \frac{1}{k_1}, k) - E(\sin^{-1} \frac{1}{k_1}, k)\right] \tag{4.72}$$

$$0 + j\,a = C_2 + \frac{2C_1}{k} \left[\left(1 - k^2/k_1^2\right) F(\sin^{-1} \frac{1}{k}, k) - E(\sin^{-1} \frac{1}{k}, k)\right] \tag{4.73}$$

$$-\frac{d}{2} + j\,0 = C_2 + \frac{2C_1}{k} \left[\left(1 - k^2/k_1^2\right) F(\sin^{-1} \frac{-1}{k_d}, k) - E(\sin^{-1} \frac{-1}{k_d}, k)\right] \tag{4.74}$$

TABLE 4.3
Mapping of Various Points between w-Plane and z-Plane as Shown in Figure 4.33

Point	w	z	β	k_i/π
A	$\rightarrow -\infty$	$\rightarrow -\infty$		
B	$0 + j\,0$	0	0	$\pi/2$
C	$0 + ja$	1	$\pi/2$	$\pi/2$
D	$-\frac{l}{2} + ja$	$\frac{1}{k_1^2}$	$\sin^{-1}(1/k_1)$	$-\pi$
E	$0 + j\,a$	$\frac{1}{k^2}$	$\sin^{-1}(1/k)$	$\pi/2$
F	$-\frac{d}{2} + j\,0$	$\frac{-1}{k_d^2}$	$\sin^{-1}(-1/k_d)$	

We recall that $F(0, k) = E(0, k) = 0$; hence, $C_2 = 0$ in all the above equations. Likewise, the incomplete elliptic integrals for the argument $\pi/2$ become complete, which we write as $F(\pi/2, k) = K(k)$, $E(\pi/2, k) = E(k)$. The second equation above allows us to write

$$C_1 = \frac{-jk}{4}\left[\frac{1}{\left(1 - \frac{k^2}{k_1^2}\right)K(k) - E(k)}\right] \tag{4.75}$$

Next, we can carry out another transformation from z-plane to r-plane to make the coplanar plates equal and symmetrical. The following simple bilinear transformation $z = \frac{r+h_1}{h_2 r + h_3}$ achieves this. This is shown in Figure 4.34. Note that E and F are taken to the points $\pm 1/k_2$, respectively.

The three constants h_1, h_2, and h_3 are obtained from the correspondence: $t = \frac{-1}{k_d^2}$, $r = \frac{-1}{k_2}$, and $t = 0$ $r = -1$, and third, $t = 1$ $r = 1$. The mapping equations are given by

$$z = \frac{r + 1}{(2 + a_0)r - a_0} \tag{4.76}$$

$$r = \frac{a_0 t + 1}{(2 + a_0)t - 1} \tag{4.77}$$

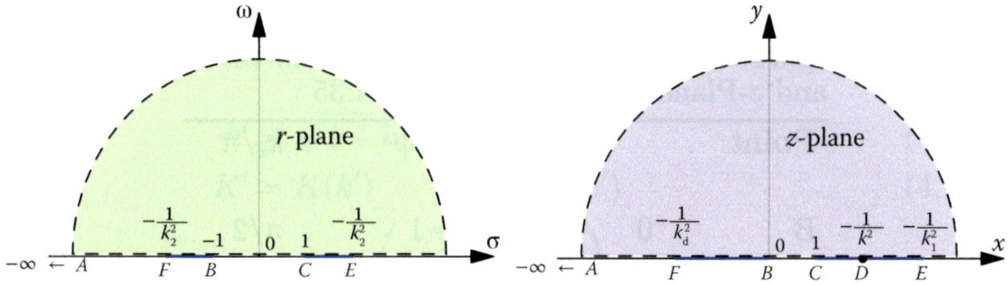

FIGURE 4.34
Mapping between r- and z-planes.

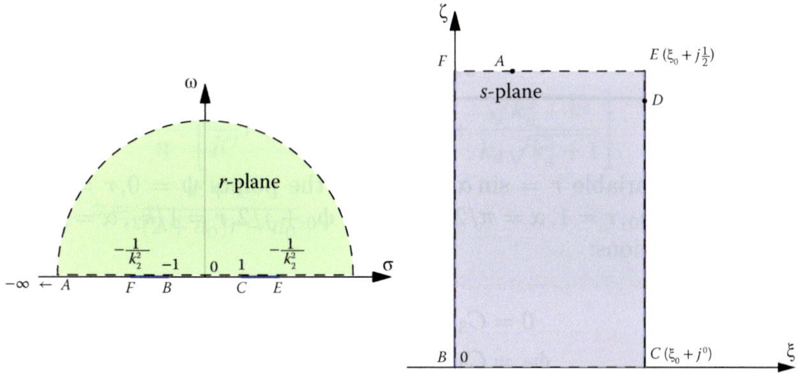

FIGURE 4.35
Mapping between r- and s-planes.

$$a_0 = \frac{k^2(1+k_2)-2}{1-k_2} \tag{4.78}$$

$$k_2 = \frac{\left[2+k_d^2-k^2+2\sqrt{(1-k^2)(1+k_d^2)}\right]}{k^2+k_d^2} \tag{4.79}$$

Last, we can apply SCT from r-plane to another say, $\psi = \phi + j\xi$ plane to obtain the familiar parallel plate geometry. This is shown in Figure 4.35.

The mapping between various points in the r-plane and the ψ-plane is shown in Table 4.4.

The mapping function is given by

$$\psi = C_3 + C_4 \int^r (r+1)^{-1/2}(r-1)^{-1/2}(r-\frac{1}{k_2})^{-1/2}(r+\frac{1}{k_2})^{-1/2}\mathrm{d}r \tag{4.80}$$

$$= C_3 + C_4 \int^r \frac{\mathrm{d}r}{\sqrt{(1-r^2)(\frac{1}{k_2^2}-r^2)}} \tag{4.81}$$

$$= C_3 + C_4 k_2 \int^r \frac{\mathrm{d}r}{\sqrt{(1-r^2)(1-k_2^2 r^2)}} \tag{4.82}$$

$$= C_3 + C_4 k_2 F(\sin^{-1} r, k_2) \tag{4.83}$$

where $K(m)$ and $K'(m)$ are the complete elliptic integrals of modulus m complementary modulus $\sqrt{(1-m^2)}$, respectively. Further U_0 and V_0 are given by

$$U_0 = 1 \tag{4.120}$$

$$V_0 = \frac{K'(m)}{K(m)} \tag{4.121}$$

Now from the w-plane, we find that the plate separation is V_0 and width is 2 units. The capacitance per unit length of the given strip line with the elliptic center conductor is twice that of the parallel plate geometry in the w-plane. Thus,

$$\mathscr{C} = \frac{4\epsilon_0\epsilon_r}{V_0} \tag{4.122}$$

$$= \frac{4\epsilon_0\epsilon_r K(m)}{K'(m)} \tag{4.123}$$

The characteristic impedance can be written as

$$Z_0 = \frac{30\pi}{\sqrt{\epsilon_r}} \frac{K'(m)}{K(m)} \tag{4.124}$$

From Equation 4.124, we note that the characteristic impedance is fixed for a given value of m. Also from Equations 4.114 and 4.115 even p and q are fixed. This in turn implies that for a fixed value of Z_0, Equation 4.116 represents a straight line in the $(r_1/b) - (r_2/b)$ plane with p and q as the horizontal and vertical intercepts on the axes. Also each fixed value of the ratio k defined as $\frac{r_1}{r_2}$ represents a straight line from the origin in the $(r_1/b) - (r_2/b)$ plane. These two families of lines are depicted in Figure 4.39.

An interesting point to appreciate is that the foregoing analysis can be specialized to the cases of horizontal, vertical, or even cylindrical strip by making $r_1 \to 0$, $r_2 \to 0$, or $r_1 = r_2$, respectively. In each case let D in general denote the width of the strip, which for the horizontal case is $2r_2$ while $r_1 = 0$, $\lambda = 0$ and for the vertical case is $2r_1$ while $r_2 = 0$, $\lambda = \infty$ and for the cylindrical case is $2r_1 = 2r_2$ while $\lambda = \frac{q}{p}$. Figure 4.40 shows variation of $\sqrt{\epsilon_r} Z_0$ as a function of $\frac{D}{2b}$. While for the horizontal case as along as strip width is finite, there is no scope for the strip to come in contact with the ground planes. But for vertical and cylindrical strips when $\frac{D}{2b} = 1$ the strip is short circuited by the ground planes and the characteristic impedance reaches zero. These details are clear from Figure 4.40.

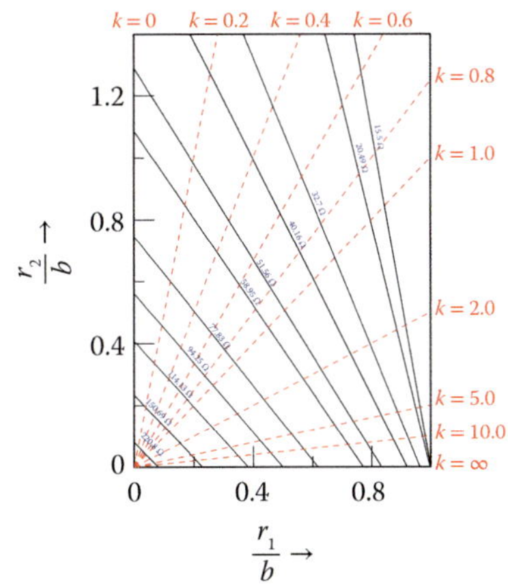

FIGURE 4.39
Family of lines showing constant Z_0 and constant k.

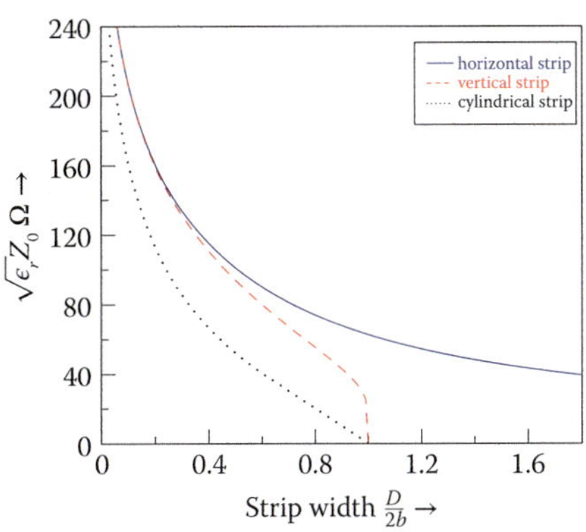

FIGURE 4.40
Showing variation in the characteristic impedance for various shapes of strips.

4.12 Conclusion

This chapter introduced the properties of analytic function of a complex variable and based on the fact that its real and imaginary parts satisfy the Laplace equation, showed how certain two-dimensional problems can be characterized by a complex potential function. Its real and imaginary parts can be used to represent electrostatic potential and flux functions. Also the idea of conformal mapping is discussed. We have shown how the two-dimensional cross-sectional geometry of a given structure can be mapped into an equivalent form that is easy for analysis. After illustrating this method through several examples, we have also introduced the Scharz–Christoffel transformation that is particularly convenient in the analysis of planar structures like strip lines.

References

[1] K. J. Binns and P. J. Lawrence, *Analysis and Computation of Electric and Magnetic Field Problems*, Oxford: Pergamon Press, 1963.

[2] J. Brown, and R. Churchill, *Complex Variables and Applications*, 8th ed., New York: McGraw-Hill, 2008.

[3] R. E. Collin, *Field Theory of Guided Waves*, New York: IEEE Press, 1991.

[4] N. N. Lebedev, I. P. Skalskaya, and Y. S. Uflyand, *Worked Problems in Applied Mathematics*, New York: Dover Publications, 1965.

[5] P. Henrici, *Applied and Computational Complex Analysis,* vol. 1 Power Series, Integration, Conformal Mapping, Location of Zeros, New York: John Wiley, 1974.

[6] W. H. Hayt, Jr., "Potential solution of a Homogeneous strip-line of finite width," *IRE Trans. Microwave Theory and Tech.,* vol. 3, no. 4, pp. 16–18, July 1955.

[7] K. V. S. Rao and B. N. Das, "Stripline using an oval-shaped center conductor between grounded planes," *IEEE Proc.*, vol. 129, pt. H, no. 6, pp. 366–368, Dec. 1982.

[8] F. Bowman, *Introduction to elliptic functions with applications*, Dover Publications, New York, 1961.

Problems

1. A uniform line charge of density ρ_L C/m lies along $x = a$, $y = b$, $a > 0$, $b > 0$, while the planes $\varphi = 0$ and $\varphi = \pi/2$ joined along the zz-axis are perfect conductors. Use the conformal mapping technique to find the potential in the first quadrant $0 \leq \varphi \leq \pi/2$.

Show that rather closer to the planes, the potential is given by

$$V = \frac{\rho_L}{\pi\epsilon_0}\frac{4xyab}{(a^2+b^2)^2}$$

and that the field lines in terms of an arbitrary constant A are given by

$$A = \frac{2\rho_L}{\pi\epsilon_0}\frac{(x^2-y^2)ab}{(a^2+b^2)^2}$$

2. Two infinitely long conducting cylinders, the contours of whose cross-section are given by

$$(x^2+y^2+\ell^2)^2 - 4\ell^2x^2 - a^4 = 0$$

$$(x^2+y^2+\ell^2)^2 - 4\ell^2x^2 - b^4 = 0, \qquad b > a > \ell$$

are held at potentials of V_1 and V_2, respectively. See Figure 4.41. Use conformal mapping technique and show that the induced charge densities are given by

$$\rho_1 = \frac{V_1 - V_2}{4\pi a^2 \log(b/a)}\sqrt{x^2+y^2}$$

$$\rho_2 = \frac{V_2 - V_1}{4\pi b^2 \log(b/a)}\sqrt{x^2+y^2}$$

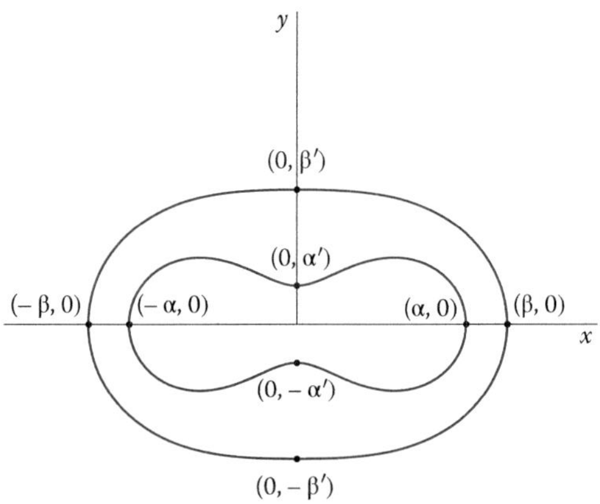

FIGURE 4.41
Cross-section of two infinitely long conducting cylinders in problem 2. $\alpha = \sqrt{a^2+\ell^2}$, $\alpha' = \sqrt{a^2-\ell^2}$, $\beta = \sqrt{b^2+\ell^2}$, $\beta' = \sqrt{b^2-\ell^2}$. The equations are given in the statement of the problem.

more elegant. In this process, we may have to involve some special functions, whose evaluation could be a challenge; but in the present day, many a software package is readily available to evaluate them.

The form of solution of a particular electromagnetic (EM) boundary value problem obtained using separation of variables technique may look different from the one obtained via transform techniques, but analytically they are identically equal according to the uniqueness theorem. If one compares variable separation method and the integral transform method, the fact that the latter derives its basis from the former would be evident.

Before we set out and introduce various transform methods, we recommend that the reader also goes through the books listed in the references [1] through [7]. In fact, many of the worked out problems and the chapter-end problems have been inspired by them, especially [1], [2], [5] and [7]. We gleaned the ideas needed for the problems and illustrated with our own figures for improving additional clarity.

5.1 The Fourier Transform

The Fourier transform of $f(x)$ denoted by $\mathcal{F}\{f(x)\}$ is defined as

$$F(k) = \frac{1}{\sqrt{2\pi}} \int_{-\infty}^{\infty} f(x) \mathrm{e}^{-jkx} \mathrm{d}x \tag{5.1}$$

The inverse Fourier transform denoted by $\mathcal{F}^{-1}\{F(k)\}$ is given by

$$f(x) = \frac{1}{\sqrt{2\pi}} \int_{-\infty}^{\infty} F(k) \mathrm{e}^{jkx} \mathrm{d}k \tag{5.2}$$

where x and k are taken on the entire real line from $-\infty$ to ∞. Note that it is possible to make alternative choices of the above definition by replacing j with $-j$. Also the coefficient $\frac{1}{\sqrt{2\pi}}$ equally attached in both Equations 5.1 and 5.2 could be replaced with $\frac{1}{2\pi}$ in one and 1 in the other. That would leave the final results unchanged.

The necessary and sufficient conditions for the existence of Equations 5.1 and 5.2 are

1. $f(x)$ should be absolutely integrable, that is, $\int_{-\infty}^{\infty} |f(x)| \, \mathrm{d}x < \infty$.
2. $f(x)$ should have finite number of finite discontinuities.
3. $f(x)$ should have finite number of minima and maxima.

Most functions that we come across in the given problem domain do satisfy the above conditions. Sometimes a function $f(t)$ is not directly transformable, but if it happens to be the limit point of a sequence of functions with a parameter, the Fourier transform is defined in the sense of the limit. Thus we include even singularity functions as they permit formal evaluation of transforms and their inverses. With that said, let us try to find the Fourier transforms of a few functions.

Example 5.1. Find the Fourier transform of the following functions from the variable z to k.

(a) $\delta(z)$
(b) $\delta(z+ut)+\delta(z-ut)$

Solution: Simply substituting the given function in the transform definition,

(a)

$$\begin{aligned}\mathcal{F}\{\delta(z)\} &= \frac{1}{\sqrt{2\pi}}\int_{-\infty}^{\infty}\delta(z)\mathrm{e}^{-jkz}\mathrm{d}z \\ &= \frac{1}{\sqrt{2\pi}}\end{aligned}$$

(b)

$$\begin{aligned}\mathcal{F}\{\delta(z+ut)+\delta(z-ut)\} &= \frac{1}{\sqrt{2\pi}}\int_{-\infty}^{\infty}\{\delta(z+ut)+\delta(z-ut)\}\mathrm{e}^{-jkz}\mathrm{d}z \\ &= \sqrt{\frac{2}{\pi}}\cos kut\end{aligned}$$

□

Example 5.2. Find the Fourier transforms of the following. The function $u(x)$ is the Heaviside function and $a>0$.

(a) $f(x)=u(x+a)-u(x-a)$
(b) $g(x)=\mathrm{e}^{-a|x|}$
(c) $h(x)=\mathrm{e}^{-ax^2}$

Solution: Direct substitution in the Equation 5.1 gives

(a)

$$\begin{aligned}\mathcal{F}\{f(x)\}=F(k) &= \frac{1}{\sqrt{2\pi}}\int_{-\infty}^{\infty}f(x)\mathrm{e}^{-jkx}\,\mathrm{d}x \\ &= \frac{1}{\sqrt{2\pi}}\int_{-a}^{a}\mathrm{e}^{-jkx}\mathrm{d}x \\ &= \frac{1}{\sqrt{2\pi}}\int_{a}^{a}(\cos kx+j\sin kx)\,\mathrm{d}x \\ &= \frac{1}{\sqrt{2\pi}}\int_{0}^{a}2\cos kx\,\mathrm{d}x \\ &= \sqrt{\frac{2}{\pi}}\left(\frac{\sin ka}{k}\right)\end{aligned}$$

(b)

$$\begin{aligned}\mathcal{F}\{g(x)\} = G(k) &= \frac{1}{\sqrt{2\pi}} \int_{-\infty}^{\infty} g(x) e^{-jkx}\, dx \\ &= \frac{1}{\sqrt{2\pi}} \int_{-\infty}^{\infty} e^{-a|x|} e^{-jkx}\, dx \\ &= \frac{1}{\sqrt{2\pi}} \int_{-\infty}^{\infty} e^{-a|x|} (\cos kx - j \sin kx)\, dx \\ &= \sqrt{\frac{2}{\pi}} \int_{0}^{\infty} e^{-ax} \cos kx\, dx \\ &= \sqrt{\frac{2}{\pi}} \left(\frac{a}{a^2 + k^2} \right)\end{aligned}$$

(c)

$$\begin{aligned}\mathcal{F}\{h(x)\} = H(k) &= \frac{1}{\sqrt{2\pi}} \int_{-\infty}^{\infty} h(x) e^{-jkx} dx \\ &= \frac{1}{\sqrt{2\pi}} \int_{-\infty}^{\infty} e^{-ax^2} e^{-jkx} dx \\ &= \frac{1}{\sqrt{2\pi}} \int_{-\infty}^{\infty} e^{-a\left\{\left(x+\frac{jk}{2a}\right)^2 + \frac{k^2}{4a^2}\right\}} dx \\ &= \sqrt{\frac{1}{2a}}\, e^{-\left(\frac{k^2}{4a}\right)}\end{aligned}$$

In deriving the above use has been made of the fact that the integral of an even function from $-a$ to a equals twice its integral from 0 to a. Often we take advantage of this fact. □

5.1.1 Properties of the Fourier Transform

The Fourier transform (and many other integral transforms) possesses several beautiful properties. We shall not list all of them but as a refresher we consider the following.

Duality: If $\mathcal{F}\{f(x)\} = F(k)$, then $\mathcal{F}\{F(x)\} = f(-k)$ or $\mathcal{F}^{-1}\{f(-k)\} = F(x)$. Using a dummy variable β, this is easily seen as follows:

$$\begin{aligned}f(x) &= \frac{1}{\sqrt{2\pi}} \int_{-\infty}^{\infty} F(k) e^{jkx} dk \\ &= \frac{1}{\sqrt{2\pi}} \int_{-\infty}^{\infty} F(\beta) e^{j\beta x} d\beta \\ \mathcal{F}\{F(x)\} &= \frac{1}{\sqrt{2\pi}} \int_{-\infty}^{\infty} F(x) e^{-jkx}\, dx \\ &= \frac{1}{\sqrt{2\pi}} \int_{-\infty}^{\infty} F(\beta) e^{-jk\beta}\, d\beta \\ &= f(-k)\end{aligned}$$

Example 5.3. Using the duality property and the Fourier transform of $e^{-a|x|}$, find the inverse Fourier transform of $e^{-a|k|}$.

Solution: We have already seen in Example 5.2 that

$$\mathcal{F}\{e^{-a|x|}\} = \sqrt{\frac{2}{\pi}}\left(\frac{a}{a^2+k^2}\right)$$

Using duality

$$\mathcal{F}^{-1}\{e^{-a|-k|}\} = \sqrt{\frac{2}{\pi}}\left(\frac{a}{a^2+x^2}\right)$$

or

$$\mathcal{F}^{-1}\{e^{-a|k|}\} = \sqrt{\frac{2}{\pi}}\left(\frac{a}{a^2+x^2}\right)$$

□

Linearity: The linearity of Fourier transform is one of the pivotal properties underlying all eigen analysis. Stated simply, it allows us to write the transform of sums as sum of the transforms. For any given constants a and b,

$$\begin{aligned}
\mathcal{F}\{af(x)+bg(x)\} &= \frac{1}{\sqrt{2\pi}}\int_{-\infty}^{\infty}(af(x)+bg(x))\,e^{-jkx}\,dx \\
&= \frac{1}{\sqrt{2\pi}}\int_{-\infty}^{\infty}af(x)e^{-jkx}\,dx + \frac{1}{\sqrt{2\pi}}\int_{-\infty}^{\infty}bg(x)e^{-jkx}\,dx \\
&= a\frac{1}{\sqrt{2\pi}}\int_{-\infty}^{\infty}f(x)e^{-jkx}\,dx + b\frac{1}{\sqrt{2\pi}}\int_{-\infty}^{\infty}g(x)e^{-jkx}\,dx \\
&= a\mathcal{F}\{f(x)\} + b\mathcal{F}\{g(x)\}
\end{aligned}$$

Shifting: This property states that the Fourier transform of a function shifted by a is product of the original transform and an exponential factor e^{-jka}.

$$\begin{aligned}
\mathcal{F}\{f(x-a)\} &= \frac{1}{\sqrt{2\pi}}\int_{-\infty}^{\infty}f(x-a)e^{-jkx}\,dx \\
&= \frac{1}{\sqrt{2\pi}}\int_{-\infty}^{\infty}f(\xi)e^{-jk(\xi+a)}\,d\xi \quad \text{with the change} \quad (x-a)=\xi \\
&= e^{-jka}\frac{1}{\sqrt{2\pi}}\int_{-\infty}^{\infty}f(\xi)e^{-jk\xi}\,d\xi \\
&= e^{-jka}\mathcal{F}\{f(x)\}
\end{aligned}$$

Scaling: This property states that the Fourier transform of a function of a new variable scaled by a is the product of the original transform in the variable on a reciprocal scale and a factor $\frac{1}{|a|}$. It also means stretching in the function domain means compression in the transform domain and vice versa.

$$
\begin{aligned}
\mathcal{F}\{f(ax)\} &= \frac{1}{\sqrt{2\pi}} \int_{-\infty}^{\infty} f(ax) \mathrm{e}^{-jkx} \, \mathrm{d}x \\
&= \frac{1}{\sqrt{2\pi}} \int_{-\infty}^{\infty} f(\xi) \mathrm{e}^{-jk(\xi/a)} \, \mathrm{d}(\xi/|a|) \quad \text{with the change } \; ax = \xi \\
&= \frac{1}{|a|} \frac{1}{\sqrt{2\pi}} \int_{-\infty}^{\infty} f(\xi) \mathrm{e}^{-j(k/a)\xi} \, \mathrm{d}\xi \\
&= \frac{1}{|a|} F(k/a)
\end{aligned}
$$

Differentiation: This property states that the Fourier transform of derivative a function is the product of the original transform and the term jk. We assume that the function $f(x)$ is piecewise smooth and continuous and that it decays to zero as $x \to \pm\infty$.

$$
\begin{aligned}
\mathcal{F}\{f'(x)\} &= \frac{1}{\sqrt{2\pi}} \int_{-\infty}^{\infty} f'(x) \mathrm{e}^{-jkx} \, \mathrm{d}x \\
&= \frac{1}{\sqrt{2\pi}} \left[f(x) \mathrm{e}^{-jkx} \Big|_{-\infty}^{\infty} + jk \int_{-\infty}^{\infty} f(x) \mathrm{e}^{-jkx} \, \mathrm{d}x \right] \\
&= jkF(k)
\end{aligned}
$$

Convolution: Convolution of two functions $f(x)$ and $g(x)$ of the real variable x denoted by $f(x) \star g(x)$ is defined as

$$
f(x) \star g(x) = \frac{1}{\sqrt{2\pi}} \int_{-\infty}^{\infty} f(\xi) g(x-\xi) \, \mathrm{d}\xi
$$

The Fourier transform of convolution of two functions $f(x)$ and $g(x)$is the product of individual transforms $F(k) \cdot G(k)$. Let us prove it below.

$$
\begin{aligned}
\mathcal{F}\{f(x) \star g(x)\} &= \frac{1}{\sqrt{2\pi}} \int_{-\infty}^{\infty} f(x) \star g(x) \, \mathrm{e}^{-jkx} \, \mathrm{d}x \\
&= \frac{1}{2\pi} \int_{-\infty}^{\infty} \int_{-\infty}^{\infty} f(\xi) g(x-\xi) \quad \mathrm{d}\xi \quad \mathrm{e}^{-jk(x-\xi)} \, \mathrm{e}^{-jk\xi} \, \mathrm{d}x \\
&= \frac{1}{2\pi} \int_{-\infty}^{\infty} f(\xi) \mathrm{e}^{-jk\xi} \left[\int_{-\infty}^{\infty} g(x-\xi) \mathrm{e}^{-jk(x-\xi)} \, \mathrm{d}(x-\xi) \right] \mathrm{d}\xi \\
&= G(k) \, \frac{1}{\sqrt{2\pi}} \int_{-\infty}^{\infty} f(\xi) \mathrm{e}^{-jk\xi} \, \mathrm{d}\xi \\
&= F(k) \cdot G(k)
\end{aligned}
$$

Parseval's Property: The Fourier transform is just a mathematical tool and cannot cause any change in the energy of the function. In topological terms, it may be viewed as an isometry that preserves norms; that is,

$$\|f(x)\| = \sqrt{\int_{-\infty}^{\infty} |f(x)|^2 \mathrm{d}x} = \|F(k)\| = \sqrt{\int_{-\infty}^{\infty} |F(k)|^2 \mathrm{d}k}$$

If $\mathcal{F}\{f(x)\} = F(k)$, it is easy to note that $f(0) = \frac{1}{\sqrt{2\pi}} \int_{-\infty}^{\infty} F(0)\ \mathrm{d}k$ and that $\mathcal{F}\left\{\overline{f(-x)}\right\} = \overline{F(k)}$ where the overline stands for complex conjugation. To prove Parseval's property, we shall employ these facts. Let us convolve $f(x)$ with $\overline{f(-x)}$ and look at the Fourier transform of the result

$$\mathcal{F}\left\{f(x) \star \overline{f(-x)}\right\} = F(k) \cdot \overline{F(k)} = |F(k)|^2$$

Using the fact that the value of the function at $x = 0$ is related to the integral of the Fourier transform,

$$f(x) \star \overline{f(-x)}\Big|_{x=0} = \frac{1}{\sqrt{2\pi}} \int_{-\infty}^{\infty} |F(k)|^2 \mathrm{d}k$$

$$\frac{1}{\sqrt{2\pi}} \int_{-\infty}^{\infty} |f(\xi)|^2\ \mathrm{d}\xi = \frac{1}{\sqrt{2\pi}} \int_{-\infty}^{\infty} |F(k)|^2 \mathrm{d}k$$

$$\|f(x)\| = \|F(k)\|$$

There are few other properties but we shall advise the reader to consult the references at the end of the chapter. Fourier transforms endowed with the beautiful mathematical properties serves as a useful tool in solving several problems in EM. There is no better way to demonstrate this than to furnish a variety of worked examples.

Example 5.4. Solve the one-dimensional wave equation with an impulsive source

$$\frac{\mathrm{d}^2\psi}{\mathrm{d}x^2} + k_0^2\psi = -\delta(x - x') \tag{5.3}$$

subject to the radiation condition $\psi(\pm\infty, x') = 0$.

Solution: Finding the Fourier transform on both the sides,

$$-k^2\Psi(k, x') + k_0^2\Psi(k, x') = -\frac{1}{\sqrt{2\pi}}\mathrm{e}^{-jkx'} \tag{5.4}$$

Hence,

$$\Phi(k, x') = \frac{1}{\sqrt{2\pi}} \frac{\mathrm{e}^{-jkx'}}{k^2 - k_0^2} \tag{5.5}$$

Inverting this again,

$$\phi(x, x') = \frac{1}{2jk_0} e^{-jk_0|x-x'|} \tag{5.6}$$

□

Example 5.5. Consider an electrostatic potential varying with x and y variables in free space. Assume that on the xz-plane, it is known to vary with x according to the function $V(x, 0) = f(x)$. Find a general expression for $V(x, y)$ for all the half-space $y > 0$. Specialize your answer to the case when the infinite strip $|x| < 1,\ y = 0$ is held at 1 V while the semi-infinite planes $y = 0,\ |x| > 1$ are grounded. That is to say, $f(x) = u(x + 1) - u(x - 1)$ where $u(x)$ is the Heaviside function. If we recall, this case corresponds to Example 4.14.

Solution: This is a typical Dirichlet problem for the half-space with a specified potential on the boundary given by $f(x)$. The potential $V(x, y)$ satisfies the Laplace's equation. Combining these two facts, we get

$$\frac{\partial^2 V}{\partial x^2} + \frac{\partial^2 V}{\partial y^2} = V_{xx} + V_{yy} = 0 \tag{5.7a}$$

$$V(x, 0) = f(x) \tag{5.7b}$$

for $-\infty < x < \infty,\ y > 0$. Let us take the Fourier transform of $V(x, y)$ in the variable x and that of $f(x)$ as

$$\mathcal{F}\{V(x, y)\} = \tilde{V}(k, y) = \frac{1}{\sqrt{2\pi}} \int_{-\infty}^{\infty} V(x, y) e^{-jkx}\ dx$$

$$\mathcal{F}\{f(x)\} = F(k) = \frac{1}{\sqrt{2\pi}} \int_{-\infty}^{\infty} f(x) e^{-jkx}\ dx$$

Using the differentiation property,

$$\mathcal{F}\{V_{xx}(x, y)\} = -k^2 \tilde{V}(k, y)$$

This changes Equation 5.7 as

$$-k^2 \tilde{V}(k, y) + \tilde{V}_{yy}(k, y) = 0$$

$$\tilde{V}(k, 0) = F(k)$$

This is just a homogeneous second order differential equation whose complimentary function is built from e^{ky} and e^{-ky}. A solution consistent with the initial condition and the fact that $V(x, y) = 0$ as $y \to \infty$ can be readily written as

$$\tilde{V}(k, y) = F(k) \cdot e^{-|k|y}$$

Let us recall from Example 5.3 that

$\mathcal{F}^{-1}\{e^{-|k|y}\} = \sqrt{\frac{2}{\pi}}\left(\frac{y}{y^2+x^2}\right)$. Since $\tilde{V}(k,y)$ is found to be product of two transforms, its inverse must be convolution of their individual inverses. Thus,

$$\begin{aligned} V(x,y) &= f(x) \star \sqrt{\frac{2}{\pi}}\left(\frac{y}{y^2+x^2}\right) \\ &= \frac{y}{\pi}\int_{-\infty}^{\infty} \frac{f(\xi)}{y^2+(\xi-x)^2}\,\mathrm{d}\xi \end{aligned}$$

For the special case of $f(x) = u(x+1)-u(x-1)$, the above expression becomes

$$\begin{aligned} V(x,y) &= \frac{y}{\pi}\int_{-1}^{1} \frac{1}{y^2+(\xi-x)^2}\,\mathrm{d}\xi \\ &= \frac{1}{\pi}\arctan\left(\frac{\xi-x}{y}\right)\Bigg|_{\xi=-1}^{1} \\ &= \frac{1}{\pi}\arctan\left(\frac{2y}{x^2+y^2-1}\right) \end{aligned}$$

This exactly tallies with the answer we got in Example 4.14. In passing, it may also be noted that if we were to include the whole space in our problem geometry, we can use symmetry considerations and use the above expression with $|y|$ in place of y. □

Next, we shall take the same geometry but pose it as a Neumann's problem by specifying surface charge density on the xz-plane.

Example 5.6. The xz-plane carries a surface charge of density $\rho_s = -2\epsilon g(x)$. The electrostatic potential $V(x,y)$ satisfies the Laplace equation for all space and for $y = 0$ it is specified by its normal derivative equal to $g(x)$. Use Fourier transform technique and solve for the potential function up to an arbitrary constant. Specialize your answer for the case $g(x) = u(x+1) - u(x-1)$.

Solution: Let us recall the fact that the discontinuity in the normal component of electric flux density must equal the surface charge density. This also implies that the normal component of electric field intensity is the ratio of surface charge density to the permittivity. So

$$E_y(x,0^+) - E_y(x,0^-) = \rho_s(x)/\epsilon \tag{5.8}$$

In terms of potential function, $E_y = -\frac{\partial V}{\partial y} = -V_y$. So we can write

$$2V_y(x,0) = -\rho_s(x)/\epsilon \tag{5.9}$$

Substituting $\rho_s = -2\epsilon g(x)$, we can formulate the problem as

$$\frac{\partial^2 V}{\partial x^2} + \frac{\partial^2 V}{\partial y^2} = V_{xx} + V_{yy} = 0 \tag{5.10a}$$

$$V_y(x,0) = g(x) \tag{5.10b}$$

Let us introduce $P(x, y) = V_y(x, y)$. Differentiating Equation 5.10a partially with respect to y, we might reformulate the problem in terms of $P(x, y)$ as

$$\frac{\partial^2 P}{\partial x^2} + \frac{\partial^2 P}{\partial y^2} = P_{xx} + P_{yy} = 0 \quad (5.11a)$$

$$P(x, 0) = g(x) \quad (5.11b)$$

which takes us back to the previous Dirichlet problem whose solution is readily found as

$$P(x, y) = \frac{y}{\pi} \int_{-\infty}^{\infty} \frac{g(\xi)}{y^2 + (\xi - x)^2} \, d\xi$$

From $P(x, y)$, we can find $V(x, y)$ by integrating with respect to y and this yields the answer up to an arbitrary constants. It may be noted that solutions of Neumann's problems are found leaving an open choice of arbitrary constant. Thus $V(x, y)$ is found as

$$\begin{aligned} V(x, y) &= \frac{1}{\pi} \int_{-\infty}^{\infty} \int^{y} \frac{\beta}{\beta^2 + (\xi - x)^2} d\beta \;\; g(\xi) \, d\xi \\ &= \frac{1}{\pi} \int_{-\infty}^{\infty} g(\xi) \log \left[\sqrt{(\xi - x)^2 + y^2} \right] d\xi \end{aligned}$$

Next for the special case of $g(x) = u(x + 1) - u(x - 1)$, we can evaluate the above integral and express the potential function as

$$V(x, y) = \frac{1}{2\pi} \{(1 - x) \log A + (1 + x) \log B + 2y \arctan C - 4\}$$

where

$$\begin{aligned} A &= \left(x^2 + y^2 - 2x + 1\right) \\ B &= \left(x^2 + y^2 + 2x + 1\right) \\ C &= \left(\frac{2y}{x^2 + y^2 - 1}\right) \end{aligned}$$

Using the relation $\mathbf{E} = -\nabla V$, we can get an expression for the electric field. Figure 5.1 shows the equipotential contours and flux lines. □

Let us see how we can solve a simple wave equation using the Fourier transform technique.

Example 5.7. The x-component of electric field in an unbounded isotropic medium of permittivity ϵ and permeability μ with an initial value of $f(z)$ and zero initial time derivative for all time and for $-\infty < z < \infty$ varies according to the scalar wave equation:

$$\frac{\partial^2 E_x}{\partial t^2} = \frac{1}{\mu\epsilon} \frac{\partial^2 E_x}{\partial z^2}$$

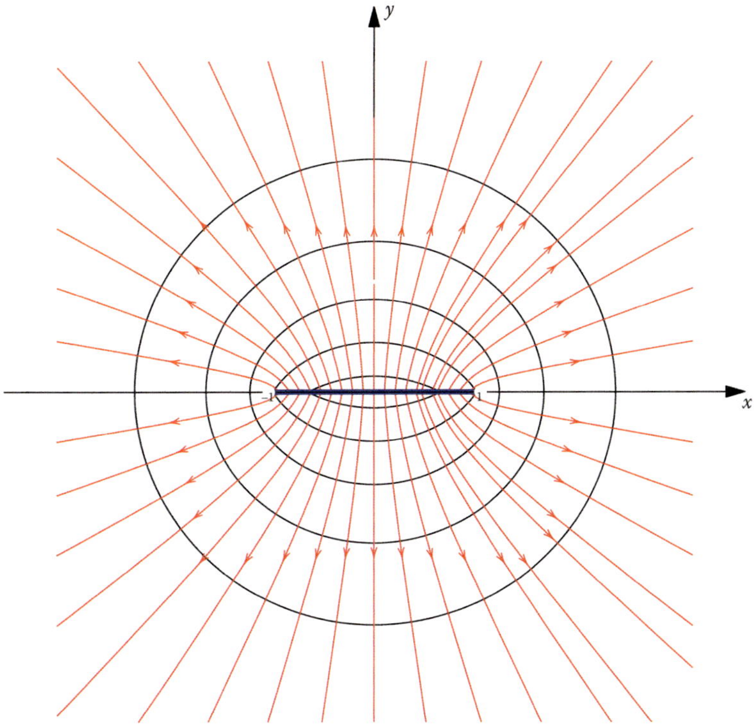

FIGURE 5.1
Equipotential and flux lines in case of an infinite strip of uniform charge lying in the xz-plane over $-1 < x < 1$.

Show that the solution for E_x is given by

$$E_x(z,t) = \frac{1}{2}\left[f(z+ut) + f(z-ut)\right]$$

where $u = 1/\sqrt{\mu\epsilon}$.

Solution: Let us retain the time variable t as is, and take the Fourier transform of the field from the variable z to ξ denoting as $\mathcal{F}\{E_x(z,t)\} = Y(\xi,t)$. Also assume that $\mathcal{F}\{f(z)\} = F(\xi)$. Equation 5.29 now becomes

$$\begin{aligned}\frac{\partial^2 Y(\xi,t)}{\partial t^2} &= -\xi^2 u^2 Y(\xi,t)\\ \frac{\partial^2 Y(\xi,t)}{\partial t^2} + \xi^2 u^2 Y(\xi,t) &= 0\end{aligned}$$

whose general solution is of the form $A(\xi)\cos(\xi ut)+B(\xi)\sin(\xi ut)$. The initial conditions

$$\begin{aligned} E_x(z,0) &= f(z) \\ \frac{\partial E_x}{\partial t} &= 0 \end{aligned}$$

in the transform domain appear as follows:

$$\begin{aligned} Y(\xi,0) &= F(\xi) \\ \left.\frac{\partial Y(\xi,t)}{\partial t}\right|_{t=0} &= 0 \end{aligned}$$

Solving for $A(\xi)$ and $B(\xi)$ subject to the initial conditions, we get

$$Y(\xi,t) = F(\xi)\cdot\cos\xi ut$$

whose inversion (using convolution property) yields

$$\begin{aligned} y(z,t) &= f(z)\star\left\{\sqrt{\frac{\pi}{2}}\left(\delta(z+ut)+\delta(z-ut)\right)\right\} \\ &= \frac{1}{2}\left[f(z+ut)+f(z-ut)\right] \end{aligned}$$

□

Example 5.8. An axially symmetric arrangement of conductors consists of a cylinder $\rho = a, \quad |z| < \ell$ with the potential $\phi(a,z) = V_0$, and two semi-infinite cylinders $\rho = a, \quad |z| > \ell$ at zero potential. Use the Fourier transform technique and solve the Laplace equation to find potential function everywhere.

Solution: The arrangement of conductors is depicted in Figure 5.2 in the bottom portion. The Laplace equation and the given boundary conditions are given by

$$\begin{aligned} \frac{1}{\rho}\frac{\partial}{\partial\rho}\left(\rho\frac{\partial\phi}{\partial\rho}\right)+\frac{\partial^2\phi}{\partial z^2} &= 0 \\ \phi(a,z) &= \begin{cases} V_0 & |z|<\ell \\ 0 & |z|>\ell \end{cases} \end{aligned}$$

Let us now define the Fourier transform of $\phi(\rho,z)$ from the variable z to the transform variable ξ as $\mathcal{F}\{\phi(\rho,z)\} = \Phi(\rho,\xi)$. The formulation can now be posed as

$$\begin{aligned} \frac{1}{\rho}\frac{d}{d\rho}\left(\rho\frac{d\Phi}{d\rho}\right)-\xi^2\Phi(\rho,\xi) &= 0 \\ \Phi(a,\xi) &= \sqrt{\frac{2}{\pi}}\frac{V_0\sin\xi\ell}{\xi} \end{aligned}$$

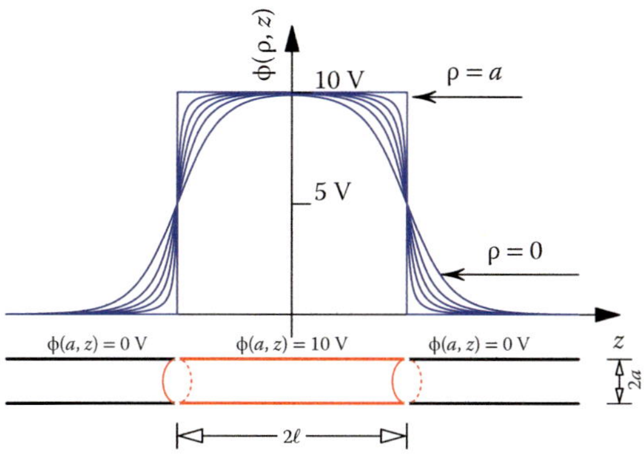

FIGURE 5.2
For $V_0 = 10$ V, $\ell = 4$ and $a = 2$, profile of the electric potential over $-10 < z < 10$ for various values of ρ from 0 to a.

a general solution of which can be written as

$$\Phi(\rho, \xi) = A(\xi)I_0(\xi\rho) + B(\xi)K_0(\xi\rho)$$

where I_0 and K_0, respectively, are the modified Bessel functions of first and second kinds of order zero; the latter being also known as the *Macdonald's function*. We discard the second part consisting of the Macdonald's function to have finite answer for $\rho = 0$. Using the boundary conditions, we get

$$\Phi(\rho, \xi) = \sqrt{\frac{2}{\pi}} \frac{V_0 \sin \xi\ell}{\xi} \frac{I_0(\xi\rho)}{I_0(\xi a)}$$

Writing the inverse, while recognizing the even symmetry, we get

$$\phi(\rho, z) = \frac{2V_0}{\pi} \int_0^\infty \frac{\cos \xi z \sin \xi\ell}{\xi} \frac{I_0(\xi\rho)}{I_0(\xi a)} d\xi$$

The above integral can be numerically computed. For illustration, we choose $V_0 = 10$ V, $a = 2$, $\ell = 4$ and for five distinct values of ρ from 0 to a that are logarithmically spaced, over the range $-10 < z < 10$, the potential is computed and plotted in Figure 5.2.

□

Example 5.9. Use the Fourier transform technique to find an expression for the potential $\phi(\rho, z)$ inside a grounded cylinder $\rho = a$ with a point charge Q lying at the origin.

Solution: In the absence of the cylinder, the potential due to the point charge is well-known:

$$\phi(\rho, z) = \frac{Q}{4\pi\epsilon_0\sqrt{\rho^2 + z^2}}$$

But because of the cylinder, we need to add a term $\psi(\rho, z)$ in such a way that the total potential evaluated on the cylinder is zero. Thus we write

$$\begin{aligned}\phi(\rho, z) &= \frac{Q}{4\pi\epsilon_0\sqrt{\rho^2+z^2}} + \psi(\rho, z)\\ \phi(a, z) = 0 &= \frac{Q}{4\pi\epsilon_0\sqrt{\rho^2+a^2}} + \psi(\rho, a)\\ \Rightarrow \psi(\rho, a) &= -\frac{Q}{4\pi\epsilon_0\sqrt{\rho^2+a^2}}\end{aligned}$$

Here $\psi(\rho, z)$ satisfies the Laplace equation, which in terms of the Fourier transform,

$$\frac{1}{\rho}\frac{\mathrm{d}}{\mathrm{d}\rho}\left(\rho\frac{\mathrm{d}\Psi}{\mathrm{d}\rho}\right) - \xi^2\Psi(\rho, \xi) = 0$$

$$\Psi(a, \xi) = \frac{-Q}{4\pi\epsilon_0}\frac{1}{\sqrt{2\pi}}\int_{-\infty}^{\infty}\frac{e^{-j\xi z}\mathrm{d}z}{\sqrt{\rho^2+z^2}} = -\frac{Q}{4\pi\epsilon_0}\sqrt{\frac{2}{\pi}}K_0(\xi a)$$

We can now write down the physically possible solution for the differential equation, once again discarding the Macdonald's function in the solution to have a finite answer for $\rho = 0$ as

$$\Psi(\rho, \xi) = -\frac{Q}{4\pi\epsilon_0}\sqrt{\frac{2}{\pi}}K_0(\xi a)\frac{I_0(\xi\rho)}{I_0(\xi a)}$$

Inverting this and putting in the expression of the total potential we get,

$$\phi(\rho, z) = \frac{Q}{4\pi\epsilon_0\sqrt{\rho^2+z^2}} - \frac{Q}{2\pi^2\epsilon_0}\int_0^{\infty}\frac{K_0(\xi a)I_0(\xi\rho)}{I_0(\xi a)}\cos\xi z\,\mathrm{d}\xi$$

Again, we numerically compute the potential as a function of ρ for the values: $Q = 4\pi\epsilon_0$, $a = 1$ at two different values of z. Figure 5.3 depicts this. Potential is also calculated for an isolated point charge and is shown for comparison.

□

Example 5.10. An infinite sheet coinciding with the xy-plane carries a time harmonic current of surface current density $J_s\mathbf{a_x}$ as shown in Figure 5.4. Use Fourier transform technique to get the radiated fields.

Solution: Let the current density be $\mathbf{J}$ and the magnetic vector potential be $\mathbf{A}$. With the choice $\nabla \times \mathbf{A} = \mathbf{B}$, the electric field is related to $\mathbf{A}$ as

$$\mathbf{E} = -j\omega\mathbf{A} + \frac{1}{j\omega\mu\epsilon}\nabla(\nabla\cdot\mathbf{A})$$

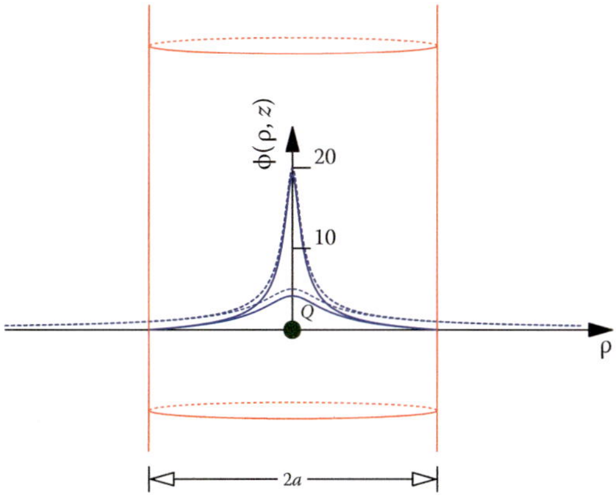

FIGURE 5.3
Variation of potential as a function of ρ for $z = 0.05$ and $z = 0.2$ with the conducting cylinder present (solid) and without it (dashed).

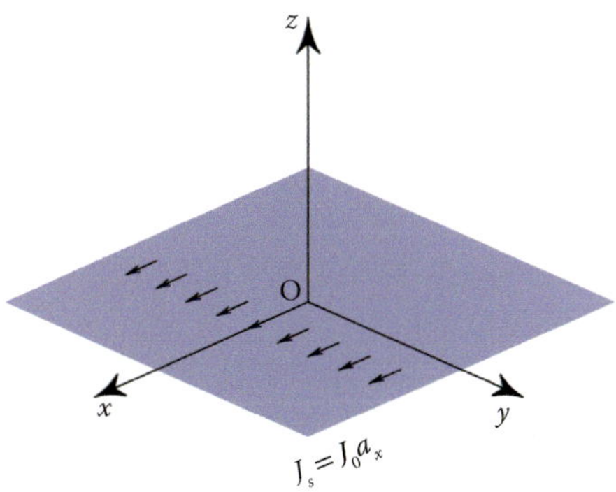

FIGURE 5.4
An infinite sheet in the xy-plane carrying a uniform surface current and radiating fields.

For the given sheet current the density is $\mathbf{J} = J_s \delta(z)\mathbf{a_x}$ and $\mathbf{A} = A_x \mathbf{a_x}$. The Helmholtz equation satisfied by $\mathbf{A}$ is

$$(\nabla^2 + k^2)\mathbf{A} = -\mu J_s \delta(z)\mathbf{a_x}$$

Since there is no variation with respect to x and y variables,

$$\frac{\mathrm{d}^2 A_x}{\mathrm{d}z^2} + k^2 A_x = -\mu J_s \delta(z)$$

Taking the Fourier transform with respect to the variable z and denoting $\mathcal{F}\{A_x(z,k)\} = \tilde{A}_x(\xi,k)$, and noting that the delta function has a transform of $\frac{1}{\sqrt{2\pi}}$,

$$\begin{aligned}
(-\xi^2 + k^2)\tilde{A}_x &= -\mu J_s \frac{1}{\sqrt{2\pi}} \\
\tilde{A}_x &= -\frac{1}{\sqrt{2\pi}}\mu J_s \frac{1}{(j^2\xi^2 - j^2k^2)} \\
&= \frac{1}{\sqrt{2\pi}}\mu J_s \frac{1}{2jk}\left(\frac{1}{j\xi + jk} - \frac{1}{j\xi - jk}\right) \\
\Rightarrow & \\
A_x(z,k) &= -\frac{j\mu J_s}{2k}\mathrm{e}^{-jk|z|}
\end{aligned}$$

Since A_x is not a function of x, $\nabla \cdot \mathbf{A} = 0$ and hence,

$$\begin{aligned}
E_x &= -j\omega A_x \\
&= -\frac{\omega\mu J_s}{2k}\mathrm{e}^{-jk|z|} \\
&= -\frac{\eta J_s}{2}\mathrm{e}^{-jk|z|}
\end{aligned}$$

From the plane wave relation $\mathbf{H} = \frac{\hat{k}\times\mathbf{E}}{\eta}$, where $\hat{k}$ is a unit vector in the direction of propagation, we get

$$H_y = -\frac{J_s}{2}\mathrm{e}^{-jk|z|}$$

We also note that the magnetic field which is tangential to the sheet is discontinuous across sheet by a value J_s as expected.

□

5.2 The Fourier Sine and Cosine Transforms

With a restriction on the domain of definition of Fourier transform we can conceive of Fourier Sine and Cosine transforms. These are particularly suitable for semi-infinite regions. We shall give the definitions and properties of these transforms, and illustrate their applications using examples.

Definitions: Suppose $f(x)$ is originally defined over $0 \leq x < \infty$ but it is extended over the domain $-\infty < x < \infty$ as an odd function, then the Fourier

Sine transform and its inverse are defined as

$$\begin{aligned}\mathcal{F}_s\{f(x)\} &= F_s(k) = \sqrt{\frac{2}{\pi}}\int_0^\infty f(x)\sin kx\,\mathrm{d}x \\ \mathcal{F}_s^{-1}\{F_s(k)\} &= f(x) = \sqrt{\frac{2}{\pi}}\int_0^\infty F_s(k)\sin kx\,\mathrm{d}k\end{aligned}$$

Likewise, suppose $f(x)$ is originally defined over $0 \le x < \infty$ but it is extended over the domain $-\infty < x < \infty$ as an even function, then the Fourier Cosine transform and its inverse are defined as

$$\begin{aligned}\mathcal{F}_c\{f(x)\} &= F_c(k) = \sqrt{\frac{2}{\pi}}\int_0^\infty f(x)\sin kx\,\mathrm{d}x \\ \mathcal{F}_c^{-1}\{F_c(k)\} &= f(x) = \sqrt{\frac{2}{\pi}}\int_0^\infty F_s(k)\sin kx\,\mathrm{d}k\end{aligned}$$

We note that when $f(x)$ is real, $\mathcal{F}\{f(x)\} = \mathcal{F}_c\{f(x)\}$ and $\mathcal{F}\{f_o(x)\} = -j\mathcal{F}_s\{f(x)\}$ where $f_e(x)$ and $f_o(x)$ are even and odd extensions of $f(x)$, respectively.

Example 5.11. Consider the half-space $0 \le x \le \infty$ and solve

$$\frac{\mathrm{d}^2\psi}{\mathrm{d}x^2} + k_0^2\psi = -\delta(x - x') \tag{5.12}$$

subject to the terminal and radiation conditions, respectively, given by

$$\psi(0, x') = \psi(\infty, x') = 0 \tag{5.13}$$

Solution: Finding Fourier Sine transform on both the sides,

$$-k^2\Psi_s(k, x') + k_0^2\Psi_s(k, x') = -\sqrt{\frac{2}{\pi}}\sin kx' \tag{5.14}$$

Hence,

$$\Psi_s(k, x') = \sqrt{\frac{2}{\pi}}\frac{\sin kx'}{k^2 - k_0^2} \tag{5.15}$$

Inverting this,

$$\psi(x, x') = \begin{cases}\frac{\sin k_0 x}{k_0}\mathrm{e}^{-jk_0 x'} & \text{for} \quad 0 \le x \le x' \\ \frac{\sin k_0 x'}{k_0}\mathrm{e}^{-jk_0 x} & \text{for} \quad x' \le x \le \infty\end{cases} \tag{5.16}$$

□

Example 5.12. Consider the the semi-infinite strip region $0 < x < \infty, 0 < y < b$ in the xy-plane shown in Figure 5.5 bounded by grounded conductors at the left and top boundaries while the bottom boundary is maintained at a potential of $f(x)$. Find an expression for the electrostatic potential $\phi(x, y)$ everywhere inside.

Solution: The potential inside the region satisfies the Laplace equation and the boundary conditions:

$$\begin{aligned}
\frac{\partial^2\phi}{\partial x^2} + \frac{\partial^2\phi}{\partial y^2} &= 0, && 0 < x < \infty, \quad 0 < y < b \\
\phi(x, 0) &= f(x), && 0 < x < \infty \\
\phi(0, y) &= 0, && 0 < y < b \\
\phi(x, b) &= 0, && 0 < x < \infty
\end{aligned}$$

Let us define $\mathcal{F}_s\{f(x)\} = F_s(k)$ and $\mathcal{F}_s\{\phi(x, y)\} = \Phi_s(k, y)$. In terms of the transforms the Laplace's equation becomes

$$-k^2\Phi_s(k, y) + \frac{\partial^2\Phi_s(k, y)}{\partial y^2} = 0$$

A solution for this can be readily written as

$$\Phi_s(k, y) = A(k)\cosh ky + B(k)\sinh ky$$

Using the boundary conditions

$$\begin{aligned}
\Phi_s(k, 0) = F_s(k) &= A(k) \\
\Phi_s(k, b) = 0 &= A(k)\cosh kb + B(k)\sinh kb
\end{aligned}$$

We can write,

$$\Phi_s(k, y) = F_s(k) \cdot \frac{\sinh k(b - y)}{\sinh kb}$$

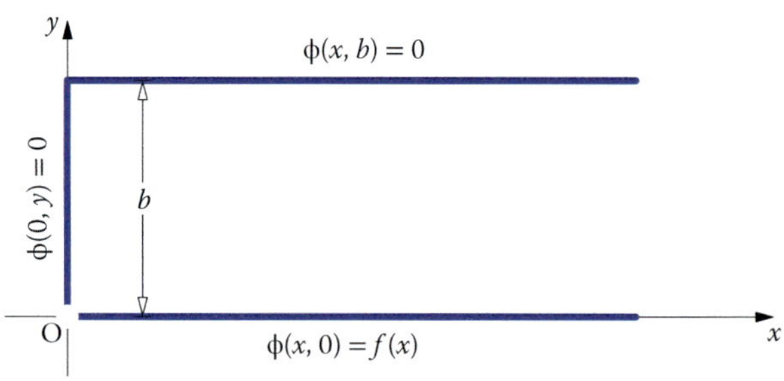

FIGURE 5.5
Boundaries for the semi-infinite region.

We can view the above as a product of two transforms

$$\begin{aligned} f(x) &\Leftrightarrow F_s(k) \\ \sqrt{\frac{2}{\pi}} \int_0^\infty \frac{\sinh k(b-y)}{\sinh kb} \sin kx dk &\Leftrightarrow \frac{\sinh k(b-y)}{\sinh kb} \\ \phi(x,y) &\Leftrightarrow F_s(k) \cdot \frac{\sinh k(b-y)}{\sinh kb} \end{aligned}$$

Although convolution property comes to mind to invert this, such an endeavor involves (i) taking the inverse Cosine transform of the second term (ii) convolving the odd extension of the first function with the even extension of the second, (iii) considering only the right (causal) half of the result, and (iv) last attaching a factor of half. Instead, it is better to take the inverse directly as

$$\begin{aligned} \phi(x,y) &= \sqrt{\frac{2}{\pi}} \int_0^\infty F_s(k) \cdot \frac{\sinh k(b-y)}{\sinh kb} \sin kx \, dk \\ &= \frac{2}{\pi} \int_0^\infty \int_0^\infty f(\xi) \frac{\sinh k(b-y)}{\sinh kb} \sin kx \sin k\xi \, dk \, d\xi \end{aligned}$$

□

Example 5.13. In the $z = 0$ plane, the electrostatic potential along the positive x-axis is fixed at zero. A uniform x-directed static electric field of E_0 is established on the y-axis over the interval $0 < y < b$. Find an expression for the potential for all $0 < x < \infty, 0 < y < \infty$. See Figure 5.6.

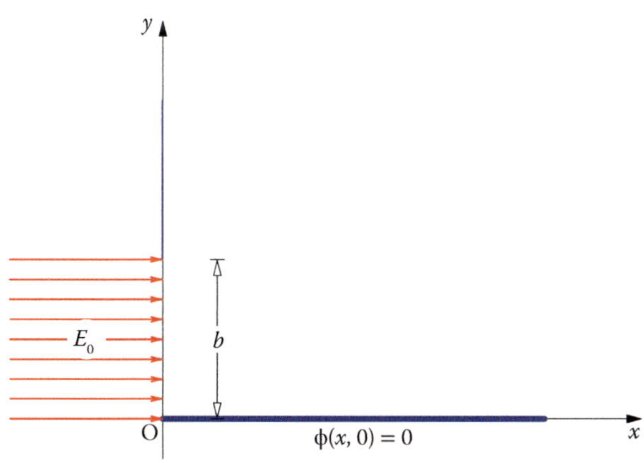

FIGURE 5.6
A uniform x-directed electric field over $0 < y < b$.

Solution: The potential $\phi(x, y)$ satisfies the Laplace's equation and the BC's given as

$$
\begin{aligned}
\frac{\partial^2\phi}{\partial x^2} + \frac{\partial^2\phi}{\partial y^2} &= 0, && 0 < x < \infty, \quad 0 < y < \infty \\
\phi(x, 0) &= 0, && 0 < x < \infty \\
\phi_x(0, y) &= -E_0, && 0 < y < b
\end{aligned}
$$

Let us retain the variable x and find the Fourier Sine transform from y to k as $\mathcal{F}_s\{\phi(x, y)\} = \Phi(x, k)$. Then the Laplace's equation becomes

$$\frac{\mathrm{d}^2\Phi(x, k)}{\mathrm{d}x^2} - k^2\Phi(x, k) = 0$$

whose general solution is of the form $A(k)\mathrm{e}^{-kx} + B(x)\mathrm{e}^{kx}$. But to have finite potential for large values of x, $B(k) = 0$. Thus,

$$\phi(x, y) = \sqrt{\frac{2}{\pi}} \int_0^\infty A(k)\mathrm{e}^{-kx} \sin ky \; \mathrm{d}k$$

Also the boundary condition on the derivative of ϕ is

$$\left.\frac{\partial\phi}{\partial x}\right|_{x=0} = \sqrt{\frac{2}{\pi}} \int_0^\infty -kA(k) \sin ky \; \mathrm{d}k$$

This implies

$$
\begin{aligned}
-kA(k) &= \sqrt{\frac{2}{\pi}} \int_0^\infty \left.\frac{\partial\phi}{\partial x}\right|_{x=0} \sin ky \; \mathrm{d}y \\
&= \sqrt{\frac{2}{\pi}} \int_0^b E_0 \sin ky \; \mathrm{d}y \\
&= \sqrt{\frac{2}{\pi}} E_0 \left(\frac{1 - \cos kb}{k}\right) \\
A(k) &= -\sqrt{\frac{2}{\pi}} E_0 \left(\frac{1 - \cos kb}{k^2}\right)
\end{aligned}
$$

Substituting for $A(k)$, we can write the solution in integral form as

$$\phi(x, y) = \frac{-2E_0}{\pi} \int_0^\infty \left(\frac{1 - \cos kb}{k^2}\right) \mathrm{e}^{-kx} \sin ky \; \mathrm{d}k$$

□

Example 5.14. A line current $i(t) = I_0 \cos \omega t$ parallel to z-axis passing through the point $(a, 0, 0)$ is inside a conducting semi-infinite box whose cross-section is shown in Figure 5.7. Find E_z radiated.

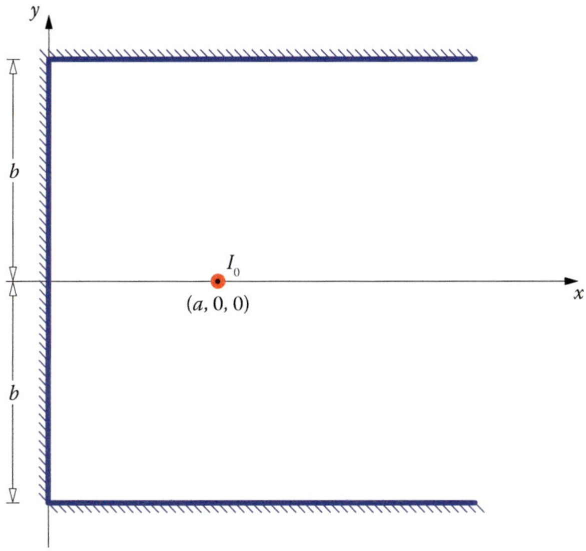

FIGURE 5.7
An infinite line source enclosed in a semi-infinite perfectly conducting trough.

Solution: Maxwell's equations in a source-included medium reveal that the electric field vector **E** satisfies the wave equation:

$$\nabla^2\mathbf{E} + k^2\mathbf{E} = j\omega\mu\mathbf{J}$$

where $k = \omega\sqrt{\mu\epsilon}$. The uniform line source of I_0 kept at $(a, 0)$ in the xy-plane can be thought of as constituting $\mathbf{J} = I_0\delta(x-a)\delta(y)\mathbf{a_z}$. Let us retain the y variable as is, but define the Fourier Sine transform from x to ξ for the z-component of the field as $\mathcal{F}_s\{E_z(x,y)\} = \Xi(\xi, y)$. Also $\mathcal{F}_s\{I_0\delta(x-a)\delta(y)\} = \sqrt{\frac{2}{\pi}}I_0\delta(y)\sin\xi a$. Then the wave equation becomes

$$\frac{\mathrm{d}^2\Xi}{\mathrm{d}y^2} - \xi^2\Xi + k^2\Xi \quad = \quad \sqrt{\frac{2}{\pi}}j\omega\mu I_0\delta(y)\sin\xi a$$

With the substitution $\sqrt{\xi^2 - k^2} = \alpha$,

$$\frac{\mathrm{d}^2\Xi}{\mathrm{d}y^2} - \alpha^2\Xi \quad = \quad \sqrt{\frac{2}{\pi}}j\omega\mu I_0\delta(y)\sin\xi a$$

whose general solution is

$$\Xi(\xi, y) \quad = \quad A(\alpha)\cosh\alpha y + B(\alpha)\sinh\alpha y + \sqrt{\frac{2}{\pi}}j\omega\mu I_0\sin\xi a\frac{\sinh\alpha|y|}{\alpha}$$

Since by symmetry it is supposed to be an even function of y, we can eliminate the second term. Applying the boundary condition $\Xi(\xi, b) = 0$, we can solve for $A(\alpha)$. Hence,

$$\begin{aligned}
\Xi(\xi, y) &= \sqrt{\frac{2}{\pi}} \frac{j\omega\mu I_0 \sin \xi a}{\alpha \cosh \alpha b} \sinh \alpha\,(|y| - b) \\
E_z(x, y) &= \frac{2}{\pi} \int_0^\infty \frac{j\omega\mu I_0 \sin \xi a}{\alpha \cosh \alpha b} \sinh \alpha\,(|y| - b) \sin \xi x \; d\xi \\
&= \frac{1}{\pi} \int_0^\infty \frac{j\omega\mu I_0}{\alpha \cosh \alpha b} \sinh \alpha\,(b - |y|) \left\{\cos \xi(x+a) - \cos \xi(x-a)\right\} \; d\xi \\
&= \frac{1}{\pi} \int_0^\infty \frac{j\omega\mu I_0}{\alpha \cosh \alpha b} \left\{\sinh \alpha b \cosh \alpha|y| - \cosh \alpha b \sinh \alpha|y|\right\} \\
&\qquad \times \left\{\cos \xi(x+a) - \cos \xi(x-a)\right\} \; d\xi
\end{aligned}$$

Let us specialize the result for the case $b \to \infty$.

$$\begin{aligned}
E_z(x, y) &= \frac{1}{\pi} \int_0^\infty \frac{j\omega\mu I_0}{\alpha} \left\{\tanh \alpha b \cosh \alpha|y| - \sinh \alpha|y|\right\} \\
&\qquad \times \left\{\cos \xi(x+a) - \cos \xi(x-a)\right\} \; d\xi \\
&= \frac{1}{\pi} \int_0^\infty \frac{j\omega\mu I_0}{\alpha} \left\{\cosh \alpha|y| - \sinh \alpha|y|\right\} \\
&\qquad \times \left\{\cos \xi(x+a) - \cos \xi(x-a)\right\} \; d\xi \\
&= \frac{\omega\mu I_0}{4} \left\{H_0^{(2)}\left(\sqrt{x^2 + a^2} + y^2\right) - H_0^{(2)}\left(\sqrt{x^2 - a^2} + y^2\right)\right\}
\end{aligned}$$

where $H_0^{(2)}(\cdot)$ is the zero order Hankel function of the second kind. □

Example 5.15. The planes $y = 0$ and $y = h$ are connected to ground and held at zero volts while an infinite uniform line charge of density ρ_L lies along $x = 0, y = a$. See Figure 5.8. Solve the Poisson equation and solve for the electrostatic potential in the regions $0 < y < a$ and $a < y < h$. After doing that, develop an expression for the charge induced on the bottom plate, take the special case when $h \to \infty$ and glide your way toward getting the result that exactly tallies with what you get from method of images.

Solution: Since there is no variation in the z-direction, we treat the electrostatic potential as being a function of x and y and denote it as $\phi(x, y)$. In the x-direction the problem has even symmetry around $x = 0$. The Poisson equation for the potential $\phi(x, y)$ and the boundary conditions are

$$\begin{aligned}
\frac{\partial^2 \phi}{\partial x^2} + \frac{\partial^2 \phi}{\partial y^2} &= -\frac{\rho_L}{\epsilon} \delta(x)\delta(y - a) \\
\phi(x, 0) &= 0 \\
\phi(x, h) &= 0
\end{aligned}$$

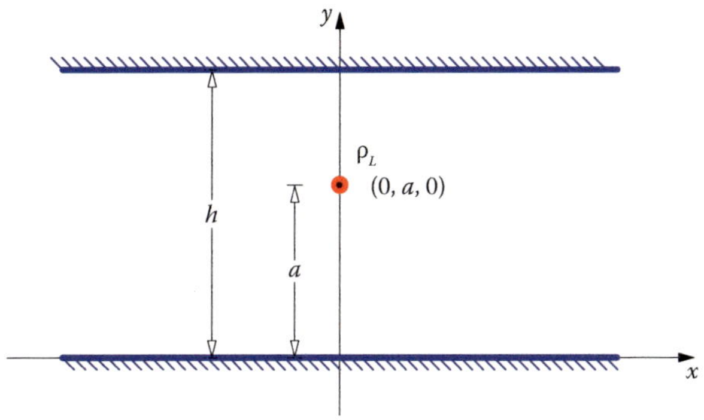

FIGURE 5.8
An infinite line charge between perfectly conducting parallel planes $y = 0$ and $y = h$.

Since there is even symmetry with respect to x, let us take the Fourier Cosine transform of $\phi(x, y)$ from the variable x to the variable ξ and denote $\mathcal{F}_c\{\phi(x, y)\} = \Phi(\xi, y)$. Remembering that there is no x-directed electric field along $x = 0$, we can formulate the problem as

$$\begin{aligned} -\xi^2\Phi + \frac{\mathrm{d}^2\Phi}{\mathrm{d}y^2} &= -\frac{\rho_L}{\epsilon}\frac{1}{\sqrt{2\pi}}\delta(y-a) \\ \Phi(x, 0) &= 0 \\ \Phi(x, h) &= 0 \end{aligned}$$

Let us designate the solution for $\Phi(\xi, y)$ in two parts:

$$\Phi(\xi, y) = \begin{cases} \Phi_1(\xi, y) & \text{for} \quad 0 < y < a \\ \Phi_2(\xi, y) & \text{for} \quad a < y < h \end{cases}$$

Since each satisfies the Laplace's equation, we can express the solutions as

$$\begin{aligned} \Phi_1(\xi, y) &= A\cosh\xi y + B\sinh\xi y \\ \Phi_2(\xi, y) &= C\cosh\xi y + D\sinh\xi y \\ \mathcal{F}_c\{E_{y1}\} &= -\frac{\partial\Phi_1}{\partial y} \\ \mathcal{F}_c\{E_{y2}\} &= -\frac{\partial\Phi_2}{\partial y} \end{aligned}$$

where A, B, C, and D are arbitrary constants that can be determined by enforcing the following boundary conditions:

$$\begin{aligned}
\Phi_1(\xi, 0) &= 0 \\
\Phi_2(\xi, h) &= 0 \\
\Phi_1(\xi, a) &= \Phi_2(\xi, a) \\
\left(\frac{\partial \Phi_1}{\partial y} - \frac{\partial \Phi_2}{\partial y}\right)\bigg|_{y=a} &= \frac{\rho_L}{\epsilon}\frac{1}{\sqrt{2\pi}}
\end{aligned}$$

Solving for the arbitrary constants A, B, C, and D, we can write

$$\begin{aligned}
\Phi_1(\xi, y) &= \frac{\rho_L}{\epsilon}\frac{1}{\sqrt{2\pi}}\frac{\sinh \xi(h-a)}{\xi \sinh \xi h}\sinh \xi y \\
\Phi_2(\xi, y) &= \frac{\rho_L}{\epsilon}\frac{1}{\sqrt{2\pi}}\frac{\sinh \xi a}{\xi \sinh \xi h}\sinh \xi(h-y)
\end{aligned}$$

Inverting the Fourier Cosine transform,

$$\begin{aligned}
\phi_1(\xi, y) &= \frac{\rho_L}{\pi\epsilon}\int_0^\infty \frac{\sinh \xi(h-a)}{\xi \sinh \xi h}\sinh \xi y \cos \xi x d\xi \\
\phi_2(\xi, y) &= \frac{\rho_L}{\pi\epsilon}\int_0^\infty \frac{\sinh \xi(h-y)}{\xi \sinh \xi h}\sinh \xi a \cos \xi x d\xi
\end{aligned}$$

The electric field can be obtained using the relation $\mathbf{E} = -\nabla\phi$. We can find the charge induced on the bottom plate using $\rho_s(x, 0) = \epsilon E_y(x, 0) = -\epsilon \left.\frac{\partial \phi_1}{\partial y}\right|_{y=0}$. Thus,

$$\begin{aligned}
\rho_s(x, 0) &= -\frac{\rho_L}{\pi}\int_0^\infty \frac{\sinh \xi(h-a)}{\sinh \xi h}\cos \xi x d\xi \\
&= -\frac{\rho_L}{\pi}\int_0^\infty (\cosh \xi a - \coth \xi h \sinh \xi a)\cos \xi x d\xi
\end{aligned}$$

Now if we take the limit $h \to \infty$ and send the top plate out of sight, the problem is simply one of a line charge above an infinite perfect electric conductor (PEC) plane. Method of images gives the answer straightway. However, if we pretend we never knew method of images, we can still get the same answer by taking the limit in the above expression. Of course, we need to remember Laplace's transform of cosine function, and the fact that $\lim \coth \theta \to 1$ as $\theta \to 0$.

$$\begin{aligned}
\rho_s(x, 0) &= -\frac{\rho_L}{\pi}\int_0^\infty (\cosh \xi a - \sinh \xi a)\cos \xi x d\xi \\
&= -\frac{\rho_L}{\pi}\int_0^\infty e^{-\xi a}\cos \xi x d\xi \\
&= -\frac{\rho_L}{\pi}\left(\frac{a}{a^2 + x^2}\right)
\end{aligned}$$

This is the exact expression one gets from the method of images. □

Example 5.16. An electron optical lens consisting of two pairs of semi-infinite parallel plates with one pair held at V_1 and another at V_2. The plates are separated by $2h$ as shown in Figure 5.9. Obtain the electrostatic potential between them. Use Fourier Sine transform.

Solution: Suppose the actual potential is $\phi(x, y)$. Let us define two dummy quantities $V_e = \frac{V_1+V_2}{2}$ and $V_o = \frac{V_2-V_1}{2}$ and a new potential function $\vartheta(x, y) = \phi(x, y) - V_e$. This permits us to solve another problem in ϑ, which has an odd symmetry so that we can use Fourier Sine transform. In the new scenario, the plates on the right are at V_o and the ones at the left are at $-V_o$. The formulation of the problem has the system

$$\begin{aligned}\frac{\partial^2\vartheta}{\partial x^2} + \frac{\partial^2\vartheta}{\partial y^2} &= 0 \\ \vartheta(x, y) &= \begin{cases} V_0 & \text{for} \quad x > 0, y = \pm h \\ -V_0 & \text{for} \quad x < 0, y = \pm h \end{cases}\end{aligned}$$

Because of odd symmetry, $\vartheta(0, y) = 0$. Taking the Fourier Sine transform of ϑ as $\mathcal{F}_s\{\vartheta(x, y)\} = \Theta(\xi, y)$, and noting $\mathcal{F}_s\{\vartheta(x, \pm h)\} = \sqrt{\frac{2}{\pi}}\frac{V_o}{\xi}$, we can write

$$\begin{aligned}-\xi^2\Theta + \Theta'' &= 0 \\ \Theta(\xi, \pm h) &= \sqrt{\frac{2}{\pi}}\frac{V_o}{\xi}\end{aligned}$$

whose solution is

$$\Theta(\xi, y) = \frac{\cosh \xi y}{\cosh \xi h}\left(\sqrt{\frac{2}{\pi}}\frac{V_o}{\xi}\right)$$

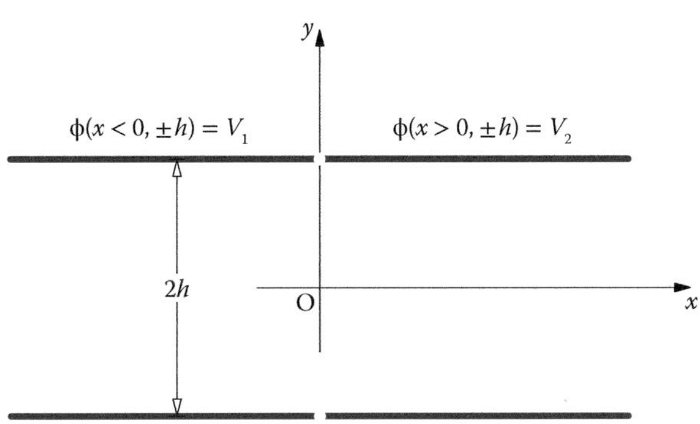

FIGURE 5.9
An electron-optical lense.

Inverting we get

$$\begin{aligned}\vartheta(x,y) &= \sqrt{\frac{2}{\pi}}\int_0^\infty \frac{\cosh\xi y}{\cosh\xi h}\left(\sqrt{\frac{2}{\pi}}\frac{V_o}{\xi}\right)\sin\xi x \mathrm{d}\xi \\ &= \frac{2V_o}{\pi}\int_0^\infty \frac{\cosh\xi y}{\cosh\xi h}\left(\frac{\sin\xi x}{\xi}\right)\mathrm{d}\xi\end{aligned}$$

Now restoring our original potential function,

$$\phi(x,y) = V_e + \frac{2V_o}{\pi}\int_0^\infty \frac{\cosh\xi y}{\cosh\xi h}\left(\frac{\sin\xi x}{\xi}\right)\mathrm{d}\xi$$

□

Example 5.17. A uniform line charge of linear density ρ_L C/m lies along the line $x = 0, y = a$ while xz-plane is a grounded PEC. Find **E** for $y > 0$ and ρ_S on the PEC.

Solution: If we represent the line charge as having a volume charge density of $\rho_v(x,y) = \rho_L(x,y)\delta(x)\delta(y-a)$ C/m^3, the potential $\phi(x,y)$ satisfies:

$$\frac{\partial^2\phi}{\partial x^2} + \frac{\partial^2\phi}{\partial y^2} = -\frac{\rho_L}{\epsilon}\delta(x)\delta(y-a)$$

Suppose the Fourier Cosine transform of $\phi(x,y)$ is denoted by $\mathcal{F}_c\{\phi(x,y)\} = \Phi(k,y)$. Also, the Fourier Cosine transform of $\delta(x)$ is $\frac{1}{\sqrt{2\pi}}$ (but not $\sqrt{\frac{2}{\pi}}$). Now,

$$-k^2\Phi + \frac{\mathrm{d}^2\Phi}{\mathrm{d}y^2} = -\frac{\rho_L}{\epsilon}\frac{1}{\sqrt{2\pi}}\delta(y-a)$$

The boundary conditions to be satisfied are

$$\begin{aligned}\Phi(k,0) &= 0 \\ \Phi(k,\infty) &= 0 \\ \Phi(k,a^-) &= \Phi(k,a^+) \\ \left.\frac{\mathrm{d}\Phi}{\mathrm{d}y}\right|_{y=a^-}^{y=a^+} &= -\frac{\rho_L}{\epsilon}\frac{1}{\sqrt{2\pi}}\end{aligned}$$

Solving for $\Phi(k, y)$ and inverting the same,

$$\Phi(k,y) = \begin{cases} A(k)e^{-ky} & \text{for} \quad y > a \\ B(k)\sinh ky & \text{for} \quad 0 < y < a \end{cases}$$

$$A(k) = \frac{\rho_L}{\epsilon}\frac{1}{\sqrt{2\pi}}\frac{\sinh ka}{k}$$

$$B(k) = \frac{\rho_L}{\epsilon}\frac{1}{\sqrt{2\pi}}\frac{e^{-ka}}{k}$$

$$\phi(x,y) = \begin{cases} \frac{\rho_L}{\pi\epsilon}\int_0^\infty \frac{\sinh ka}{k}e^{-ky}\cos kx\,dk & y > a \\ \frac{\rho_L}{\pi\epsilon}\int_0^\infty \frac{e^{-ka}}{k}\sinh ky\cos kx\,dk & 0 < y < a \end{cases}$$

Although the expression derived above at first sight looks different from the familiar result obtained via method of images, we can readily show that it is the same. Let us start with the expression for $y > a$ and conveniently use Laplace's transform table and properties.

$$\begin{aligned} \phi(x,y) &= \frac{\rho_L}{\pi\epsilon}\int_0^\infty \frac{\sinh ka}{k}e^{-ky}\cos kx\,dk \\ &= \frac{\rho_L}{2\pi\epsilon}\int_y^\infty \left.\frac{y}{y^2+x^2}\right|_{y+1}^{y-a} dy \\ &= \frac{\rho_L}{2\pi\epsilon}\log\left(\sqrt{\frac{(y+a)^2+x^2}{(y-a)^2+x^2}}\right) \end{aligned}$$

This agrees with the result obtained from the method of images. Suppose we take the expression for $\phi(x, y)$ for $0 < y < a$, we would get a similar answer with $(a - y)^2$ in place of $(y - a)^2$ which does not matter. Thus, we find the answer for potential in terms of logarithm is valid for both regions $0 < y < a$ and $y > a$, whereas it had to be spelled separately when written in terms of Cosine transform. The electric field intensity for $0 < y < a$ and in particular along $y = 0$ is given by

$$\begin{aligned} \mathbf{E} &= -\nabla\left\{\frac{\rho_L}{\pi\epsilon}\int_0^\infty \frac{e^{-ka}}{k}\sinh ky\cos kx\,dk\right\} \\ E_y &= -\frac{\rho_L}{\pi\epsilon}\int_0^\infty e^{-ka}\cosh ky\cos kx\,dk \\ E_y\big|_{y=0} &= -\frac{\rho_L}{\pi\epsilon}\int_0^\infty e^{-ka}\cos kx\,dk \\ &= -\frac{\rho_L}{\pi\epsilon}\left(\frac{a}{a^2+x^2}\right) \end{aligned}$$

From this, we can write the surface charge density as

$$\rho_S = -\frac{\rho_L}{\pi}\left(\frac{a}{a^2+x^2}\right)$$

This is what is expected when we use the method of images! □

5.3 The Hankel Transform

In this section, we introduce the Hankel transform. Let $f(\rho)$ be a function of radial distance $\rho \geq 0$. The Hankel transform of $f(\rho)$ of order $\nu > -1/2$, denoted by $\mathcal{H}_\nu\{f(\rho)\} = F_\nu(k)$ is defined as

$$F_\nu(k) = \int_0^\infty f(\rho) J_\nu(k\rho)\rho d\rho \tag{5.17}$$

whose inversion is given by

$$f(\rho) = \int_0^\infty F_\nu(k) J_\nu(k\rho) k dk \tag{5.18}$$

In order that Equations 5.17 and 5.18 converge, we need the following sufficient conditions:

1. $f(\rho) = O\left(\rho^{-k}\right)$, $\rho \to \infty$ for $k > 3/2$.
2. $f'(\rho)$ is piecewise continuous over a subinterval of the domain of its definition.
3. Whenever $f(\rho)$ is discontinuous at $\rho = a$, the inversion formula in Equation 5.18 converges to the mean of the left and right limits at the discontinuity. Stated otherwise, $f(a) = \frac{1}{2}\left[f(a^+) + f(a^-)\right]$.

We can develop a relation between the Fourier transform of a function with circular symmetry and its Hankel transform. Let $\mathcal{F}\{f(x,y)\} = F(\tilde{x},\tilde{y})$ be the two dimensional Fourier transform. Also let $f(x,y)$ depend only on $\rho = \sqrt{x^2+y^2}$.

$$F(\tilde{x},\tilde{y}) = \frac{1}{2\pi}\int_{-\infty}^\infty\int_{-\infty}^\infty f(\rho) e^{j(x\tilde{x}+y\tilde{y})} dx dy \tag{5.19}$$

Substituting $x = \rho\cos\varphi$, $y = \rho\sin\varphi$, $\tilde{x} = k\cos\vartheta$, $\tilde{y} = k\sin\vartheta$ and changing the manner of integration from rectangular to polar coordinates, we get

$$F(\tilde{x},\tilde{y}) = \frac{1}{2\pi}\int_0^\infty f(\rho)\rho d\rho \int_0^{2\pi} e^{j\rho k\cos(\varphi-\vartheta)} d\varphi \tag{5.20}$$

$$= \int_0^\infty f(\rho) J_0(k\rho)\rho d\rho \tag{5.21}$$

Since we found that $F(\tilde{x}, \tilde{y})$ does not depend on ϑ, we might as well write $F(\tilde{x}, \tilde{y}) = F(\rho) = \mathcal{H}_0\{f(\rho)\}$. Thus the two-dimensional Fourier transform of a function having circular symmetry can be interpreted as zero-order Hankel transform. Like the Fourier transform, Hankel transform also has interesting properties. It obeys linearity, duality, Parseval's property but does not possess shifting property. Let us review the differentiation property. The Laplacian operator in cylindrical coordinates is written as

$$\nabla^2 \equiv \frac{1}{\rho}\frac{\partial}{\partial\rho}\left(\rho\frac{\partial}{\partial\rho}\right) + \frac{1}{\rho^2}\frac{\partial^2}{\partial\varphi^2} + \frac{\partial^2}{\partial z^2} \tag{5.22}$$

In a rather generalized sense, for an order of ν, the ρ-part of this operator, also called the Bessel differential operator can be formally written as

$$\nabla_\nu^2 \equiv \frac{1}{\rho}\frac{\mathrm{d}}{\mathrm{d}\rho}\left(\rho\frac{\mathrm{d}}{\mathrm{d}\rho}\right) - \frac{\nu^2}{\rho^2} \tag{5.23}$$

The differentiation property of the Hankel transform helps us to solve boundary value problems in EM. The Hankel transform of order ν of the Bessel operator of order ν is understood in the following way:

$$\mathcal{H}_\nu\{\nabla_\nu^2 f(\rho)\} = \int_0^\infty \nabla_\nu^2 f J_\nu(k\rho)\rho\mathrm{d}\rho \tag{5.24}$$

$$= \int_0^\infty \left[\frac{1}{\rho}\frac{\mathrm{d}}{\mathrm{d}\rho}\left(\rho\frac{\mathrm{d}f}{\mathrm{d}\rho}\right) - \frac{\nu^2}{\rho^2}f\right] J_\nu(k\rho)\rho\mathrm{d}\rho \tag{5.25}$$

Integrating by parts and assuming the asymptotic vanishing nature of $f(\rho)$ and its derivatives, we get

$$\mathcal{H}_\nu\{\nabla_\nu^2 f(\rho)\} = \int_0^\infty k^2\left[J_\nu''(k\rho) + \frac{1}{k\rho}J'(k\rho) - \frac{\nu^2}{k^2\rho^2}J_\nu(k\rho)\right] f(\rho)\rho\mathrm{d}\rho \tag{5.26}$$

From the very fact of how $J_\nu(\cdot)$ satisfies its own differential equation, we can further write,

$$\mathcal{H}_\nu\{\nabla_\nu^2 f(\rho)\} = -k^2\int_0^\infty J_\nu(k\rho)f(\rho)\rho\mathrm{d}\rho \tag{5.27}$$

$$= -k^2\mathcal{H}_\nu\{f(\rho)\} \tag{5.28}$$

Thus we conclude that for the Hankel transform, the Bessel differential operator acts in an eigen manner, in the sense, ∇_ν^2 can be replaced by $-k^2$, just as $\frac{\mathrm{d}^2}{\mathrm{d}t^2}$ can be replaced by $-\omega^2$ in case of phasors! Before we illustrate the use of Hankel transform, let us consider some of its properties.

5.3.1 Properties of the Hankel Transform

The properties of the forward and the inverse transform are alike because the expressions are of the same form.

Linearity: This property states that if C_1 and C_2 are arbitrary constants,

$$\mathcal{H}_\nu\left\{C_1 f(\rho)+C_2 g(\rho)\right\}=C_1\mathcal{H}_\nu\left\{f(\rho)\right\}+C_2\mathcal{H}_\nu\left\{C_2 g(\rho)\right\} \tag{5.29a}$$

$$\mathcal{H}_\nu^{-1}\left\{C_1 F_\nu(k)+C_2 G_\nu(k)\right\}=C_1\mathcal{H}_\nu^{-1}\left\{F_\nu(k)\right\}+C_2\mathcal{H}_\nu^{-1}\left\{C_2 G_\nu(\rho)\right\} \tag{5.29b}$$

The fact of linearity comes directly from that of the integral operation.

Scaling: Supposing $f(\rho)$ and $F_\nu(k)$ constitute a Hankel transform pair, by a change of variables from ρ to $a\rho$ where $a>0$, we get

$$\begin{aligned}\mathcal{H}_\nu\left\{f(a\rho)\right\} &= \int_0^\infty \rho f(a\rho)J_\nu(kr)\mathrm{d}\rho\\ &= \frac{1}{a^2}\int_0^\infty \rho f(\rho)J_\nu\left(\frac{kr}{a}\right)\mathrm{d}\rho\end{aligned}$$

From the above we recognize that

$$\mathcal{H}_\nu\left\{f(a\rho)\right\} = \frac{1}{a^2}F\left(\frac{k}{a}\right) \tag{5.30}$$

Shifting: Shifting property in case of the Hankel transform is not convenient to use. One might attempt a formula using the Neumann–Lommel addition formula for the Bessel functions, namely,

$$J_\nu\left\{x+y\right\} = \sum_{\mu=-\infty}^{\infty} J_\mu(x)J_{\nu-\mu}(y)$$

This rather appears to be "convolution of orders" but an actual application of this for shifting property becomes unwieldy, so we generally overlook this property.

Differentiation: In developing this property we recall a recurrence relation satisfied by the Bessel functions, which is

$$J_{\nu+1}(\rho) = \frac{2\nu}{\rho}J_\nu(x)-J_{\nu-1}(x)$$

Now consider the Hankel transform of $f'(\rho)$ of order 1. We have

$$\begin{aligned}\mathcal{H}_1\left\{f'(\rho)\right\} &= \int_0^\infty \rho f'(\rho)J_1(k\rho)\mathrm{d}\rho\\ &= \rho f(\rho)J_1(k\rho)\big|_0^\infty-\int_0^\infty f(\rho)\frac{\mathrm{d}}{\mathrm{d}\rho}\left[\rho J_1(k\rho)\right]\mathrm{d}\rho\end{aligned}$$

We note that the first term vanishes at both the limits. Also it is known that

$$\frac{\mathrm{d}}{\mathrm{d}\rho}\left[\rho J_1(k\rho)\right] = k\rho J_0(k\rho)$$

This leaves us with,

$$\mathcal{H}_1\left\{f'(\rho)\right\} = -k\int_0^\infty \rho f(\rho)J_0(k\rho)\mathrm{d}\rho$$

which can be rewritten as

$$\mathcal{H}_1\left\{f'(\rho)\right\} = -k\mathcal{H}_0\left\{f(\rho)\right\} \quad (5.31)$$

In a similar fashion, we can get

$$\mathcal{H}_0\left\{\frac{1}{\rho}\frac{\mathrm{d}}{\mathrm{d}\rho}\left[\rho f(\rho)\right]\right\} = k\mathcal{H}_1\left\{f(\rho)\right\} \quad (5.32)$$

For higher orders, Equations 5.31 and 5.32 can be generalized as

$$\mathcal{H}_\nu\left\{\rho^{\nu-1}\frac{\mathrm{d}}{\mathrm{d}\rho}\left[\rho^{1-\nu}f(\rho)\right]\right\} = -k\mathcal{H}_{\nu-1}\left\{f(\rho)\right\} \quad (5.33)$$

$$\mathcal{H}_\nu\left\{\frac{1}{\rho^{\nu+1}}\frac{\mathrm{d}}{\mathrm{d}\rho}\left[\rho^{1+\nu}f(\rho)\right]\right\} = k\mathcal{H}_{\nu+1}\left\{f(\rho)\right\} \quad (5.34)$$

Parseval property: Let $F_\nu(k)$ and $G_\nu(k)$ be the Hankel transforms of $f(\rho)$ and $g(\rho)$, respectively. Parseval's property is a generalized version of the isometry that prevails in case of orthogonal transforms. It is given by

$$\int_0^\infty kF_\nu(k)G_\nu(k)\mathrm{d}k = \int_0^\infty \rho f(\rho)g(\rho)\mathrm{d}\rho \quad (5.35)$$

Example 5.18. Assuming $a > 0$, find the following Hankel transforms.

(a) $\mathcal{H}_0\left\{\frac{1}{\rho}\mathrm{e}^{-a\rho}\right\}$
(b) $\mathcal{H}_\nu\left\{\rho^{\nu-1}\mathrm{e}^{-a\rho}\right\}$
(c) $\mathcal{H}_0\left\{\mathrm{e}^{-a^2\rho^2}\right\}$
(d) $\mathcal{H}_\nu\left\{\rho^\nu\mathrm{e}^{-a^2\rho^2}\right\}$

Solution:

(a)

$$\begin{aligned}\mathcal{H}_0\left\{\frac{1}{\rho}\mathrm{e}^{-a\rho}\right\} &= \int_0^\infty \mathrm{e}^{-a\rho}J_0(k\rho)\mathrm{d}\rho \\ &= \int_0^\infty \mathrm{e}^{-a\rho}\mathrm{d}\rho\int_0^{\pi/2}\cos\left(k\rho\sin\varphi\right)\mathrm{d}\varphi \\ &= \frac{2}{\pi}\int_0^{\pi/2}\mathrm{d}\varphi\int_0^\infty \mathrm{e}^{-a\rho}\cos\left(k\rho\sin\varphi\right)\mathrm{d}\rho \\ &= \frac{2}{\pi}\int_0^{\pi/2}\frac{a\mathrm{d}\varphi}{a^2+k^2\sin^2\varphi} \\ &= \frac{1}{\sqrt{a^2+k^2}}\end{aligned}$$

(b)

$$\begin{aligned}\mathcal{H}_\nu\left\{\rho^{\nu-1}\mathrm{e}^{-a\rho}\right\} &= \int_0^\infty \rho^\nu\mathrm{e}^{-a\rho}J_\nu(k\rho)\mathrm{d}\rho \\ &= \mathcal{L}\left\{\rho^\nu J_\nu(k\rho), \rho\to a\right\} \\ &= \frac{(2k)^\nu\,\Gamma\left(\nu+\frac{1}{2}\right)}{\sqrt{\pi}\left(k^2+a^2\right)^{\nu+\frac{1}{2}}}\end{aligned}$$

(c)

$$\begin{aligned}\mathcal{H}_0\left\{\mathrm{e}^{-a^2\rho^2}\right\} &= \int_0^\infty \rho\mathrm{e}^{-a^2\rho^2}J_0(k\rho)\mathrm{d}\rho \\ &= \int_0^\infty \rho\mathrm{e}^{-a^2\rho^2}\sum_{n=0}^\infty\frac{(-1)^n(k\rho/2)^{2n}}{(n!)^2}\mathrm{d}\rho \\ &= \sum_{n=0}^\infty\frac{(-1)^n(k/2)^{2n}}{(n!)^2}\int_0^\infty \rho^{2n+1}\mathrm{e}^{-a^2\rho^2}\mathrm{d}\rho \\ &= \frac{1}{2a^2}\sum_{n=0}^\infty\frac{(-1)^n(k^2/4a^2)^n}{(n!)^2}\int_0^\infty p^n\mathrm{e}^{-p}\mathrm{d}p \\ &= \frac{1}{2a^2}\sum_{n=0}^\infty\frac{(-1)^n(k^2/4a^2)^n}{n!} \\ &= \frac{1}{2a^2}\mathrm{e}^{-k^2/4a^2}\end{aligned}$$

(d)

$$\begin{aligned}
\mathcal{H}_\nu\left\{\rho^\nu e^{-a^2\rho^2}\right\} &= \int_0^\infty \rho^{\nu+1}e^{-a^2\rho^2}J_\nu(k\rho)d\rho \\
&= \int_0^\infty \rho^{\nu+1}e^{-a^2\rho^2}\left[\sum_{n=0}^{\infty}\frac{(-1)^n(k\rho/2)^{2n+\nu}}{n!\Gamma(n+\nu+1)}\right]d\rho \\
&= \left[\sum_{n=0}^{\infty}\frac{(-1)^n(k/2)^{2n+\nu}}{n!\Gamma(n+\nu+1)}\right]\int_0^\infty e^{-a^2\rho^2}\rho^{2\nu+2n+1}d\rho \\
&= \frac{k^\nu}{(2a^2)^{\nu+1}}\left[\sum_{n=0}^{\infty}\frac{(-k^2/4a^2)}{n!}\right] \\
&= \frac{k^\nu}{(2a^2)^{\nu+1}}e^{-k^2/4a^2}
\end{aligned}$$

□

Next, let us consider some examples illustrate how the Hankel transform can be used. In most cases, we come across the Hankel transform of order zero.

Example 5.19. A circular conducting plate of unit radius lying in the xy-plane centered at the origin is held at V_0 Volt as shown in Figure 5.10. Find potential in the upper half space $z > 0$.

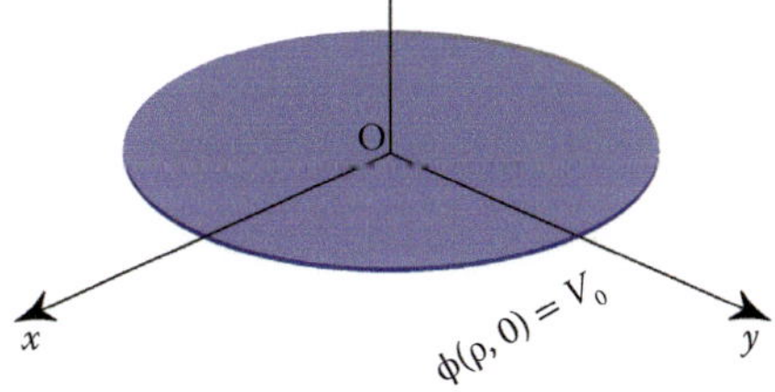

FIGURE 5.10
A circular conducting plate held at V_0.

Solution: We can use cylindrical coordinate system. There is no variation with respect to φ; so we might write the Laplace's equation and the boundary conditions as

$$
\begin{aligned}
\frac{1}{\rho}\frac{\partial}{\partial\rho}\left(\rho\frac{\partial\phi}{\partial\rho}\right)+\frac{\partial^2\phi}{\partial z^2} &= 0 \\
\phi(\rho,0) &= V_0 && 0<\rho<1 \\
\left.\frac{\partial\phi}{\partial z}\right|_{z=0} &= 0 && \rho>1 \\
\phi(\rho,\pm\infty) &= 0
\end{aligned}
$$

Let us denote the zero-order Hankel transform of the potential function from the variable ρ to k as $\mathcal{H}_0\{\phi(\rho,z)\}=\Phi(k,z)$. This translates the Laplace's equation as $-k^2\Phi(k,z)+\Phi(k,z)=0$ whose solution vanishes for $z=\infty$ is of the form $\Phi(k,z)=A(k)\mathrm{e}^{-kz}$ where $A(k)$ is to be determined from the boundary conditions. The inverse Hankel transform of this result yields

$$
\phi(\rho,z)=\int_0^\infty A(k)\mathrm{e}^{-kz}J_0(k\rho)k\mathrm{d}k
$$

Now the boundary conditions permit us to write the following:

$$
\begin{aligned}
\phi(\rho,0)=V_0 \quad &\Rightarrow \quad \int_0^\infty A(k)J_0(k\rho)k\mathrm{d}k=V_0 && 0<\rho<1 \\
\left.\frac{\partial\phi}{\partial z}\right|_{z=0}=0 \quad &\Rightarrow \quad \int_0^\infty A(k)(-k^2)J_0(k\rho)\mathrm{d}k=0 && \rho>1
\end{aligned}
$$

Let us recall two identities of the Bessel function

$$
\begin{aligned}
\int_0^\infty k^{-1}J_0(k\rho)\sin k\mathrm{d}k &= \frac{\pi}{2} && 0<\rho<1 \\
\int_0^\infty J_0(k\rho)\sin k\mathrm{d}k &= 0 && \rho>1
\end{aligned}
$$

This enables us to figure out the arbitrary coefficient $A(k)$ as

$$
\begin{aligned}
\frac{kA(k)}{V_0} &= \frac{2\sin k}{k\pi} \\
A(k) &= \frac{2V_0\sin k}{k^2\pi}
\end{aligned}
$$

Thus the sought for potential function can be written as

$$
\phi(\rho,z)=\frac{2V_0}{\pi}\int_0^\infty \frac{\sin k}{k^2}\mathrm{e}^{-kz}J_0(k\rho)k\mathrm{d}k
$$

□

Example 5.20. The xy-plane is held at a variable potential that varies only with radial distance ρ given by $f(\rho)$. Solve for the potential function from the Laplace's equation. Use Hankel transform.

Solution: In this case, the potential is not a function of the azimuthal angle φ, and we can write the Laplace's equation for $\phi(\rho, z)$ and the associated boundary conditions as

$$\begin{aligned}\frac{1}{\rho}\frac{\partial}{\partial\rho}\left(\rho\frac{\partial\phi}{\partial\rho}\right)+\frac{\partial^2\phi}{\partial z^2} &= 0\\ \phi(\rho,0) &= f(\rho)\\ \phi(\rho,\pm\infty) &= 0\end{aligned}$$

Denoting the Hankel transform of order zero $\mathcal{H}_0\{\phi(\rho,z)\} = \Phi(k,z)$ and $\mathcal{F}_H\{f(\rho)\} = F(k)$, we get

$$\begin{aligned}-k^2\Phi(k,z)+\frac{\mathrm{d}^2\Phi}{\mathrm{d}z^2} &= 0\\ \Phi(k,0) &= F(k)\\ \Phi(k,\pm\infty) &= 0\end{aligned}$$

whose solution can be readily set down as

$$\Phi(k,z) = F(k)\mathrm{e}^{-k|z|}$$

Inverting we get

$$\begin{aligned}\phi(\rho,z) &= \int_0^\infty F(k)\mathrm{e}^{-k|z|}J_0(k\rho)k\mathrm{d}k\\ &= \int_0^\infty\int_0^\infty f(\eta)J_0(\eta k)\mathrm{e}^{-k|z|}J_0(k\rho)k\eta\mathrm{d}\eta\mathrm{d}k\end{aligned}$$

□

Example 5.21. The planes $z = \pm\ell$ are grounded PECs while a point charge Q lies at $(0, 0, 0)$ at the origin as shown in Figure 5.11. Find the electrostatic potential $\phi(\rho, z)$ for $|z| < \ell$ using Hankel transforms.

Solution: We can write the potential function

$$\phi(\rho,z) = \vartheta(\rho,z)+\frac{Q}{4\pi\epsilon\sqrt{\rho^2+z^2}}$$

where $\vartheta(\rho, z)$ satisfies the Laplace equation. Let us denote the Hankel transform $\mathcal{H}_0\{\vartheta(\rho,z)\} = \Theta(k,z)$. Then the Laplace's equation in terms of $\Theta(k, z)$ becomes

$$-k^2\Theta(k,z)+\frac{\mathrm{d}^2\Theta}{\mathrm{d}z^2} = 0$$

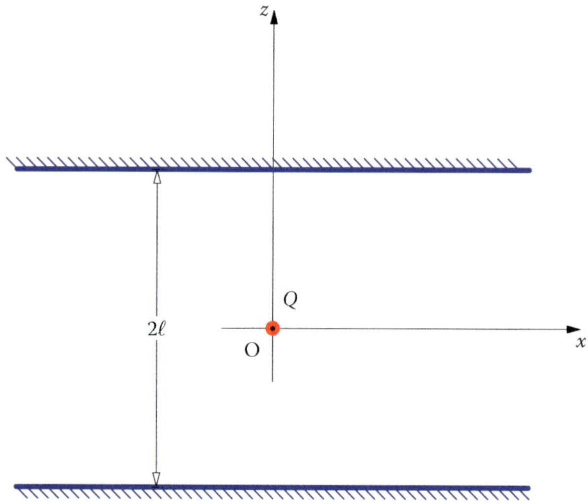

FIGURE 5.11
A point charge Q located at the origin between perfectly conducting parallel planes $z = \pm\ell$.

whose solution is of the form

$$\Theta(k, z) = A(k)\mathrm{e}^{-kz} + B(k)\mathrm{e}^{kz}$$

where $A(k)$ and $B(k)$ are arbitrary and must be determined in keeping with the boundary conditions

$$\begin{aligned}
\phi(\rho, \ell) = 0 \quad &\Rightarrow \quad A(k)\mathrm{e}^{-k\ell} + B(k)\mathrm{e}^{k\ell} + \mathcal{H}_0\left\{\frac{Q}{4\pi\epsilon\sqrt{\rho^2 + \ell^2}}\right\} = 0 \\
&\Rightarrow \quad A(k)\mathrm{e}^{-k\ell} + B(k)\mathrm{e}^{k\ell} + \frac{Q\mathrm{e}^{-k\ell}}{4\pi\epsilon k} = 0 \\
\phi(\rho, -\ell) = 0 \quad &\Rightarrow \quad A(k)\mathrm{e}^{k\ell} + B(k)\mathrm{e}^{-k\ell} + \mathcal{H}_0\left\{\frac{Q}{4\pi\epsilon\sqrt{\rho^2 + \ell^2}}\right\} = 0 \\
&\Rightarrow \quad A(k)\mathrm{e}^{k\ell} + B(k)\mathrm{e}^{-k\ell} + \frac{Qe^{-k\ell}}{4\pi\epsilon k} = 0
\end{aligned}$$

Solving for $A(k)$ and $B(k)$ and utilizing them in potential expression, we get

$$\begin{aligned}
A(k) = B(k) &= -\frac{Q}{8\pi\epsilon k}\frac{\mathrm{e}^{-k\ell}}{\cosh k\ell} \\
\Theta(k, z) &= -\frac{Q}{4\pi\epsilon k}\left\{\frac{\mathrm{e}^{-k\ell}\cosh kz}{\cosh k\ell}\right\} \\
\phi(\rho, z) &= \frac{Q}{4\pi\epsilon\sqrt{\rho^2 + z^2}} - \int_0^\infty \frac{Q}{4\pi\epsilon k}\left\{\frac{\mathrm{e}^{-k\ell}\cosh kz}{\cosh k\ell}\right\} J_0(k\rho)k\mathrm{d}k
\end{aligned}$$

A quick glance at two trivial cases when $\ell \to \infty$ and $\ell \to 0$ can be taken.

$$
\begin{aligned}
\phi(\rho, z)|_{\ell\to\infty} &= \frac{Q}{4\pi\epsilon\sqrt{\rho^2+z^2}} - 0 \\
&= \frac{Q}{4\pi\epsilon\sqrt{\rho^2+z^2}} \\
\phi(\rho, z)|_{\ell\to 0} &= \frac{Q}{4\pi\epsilon\sqrt{\rho^2+z^2}} - \int_0^\infty \frac{Q}{4\pi\epsilon k}\left\{\frac{e^{-k\ell}\cosh kz}{2}\right\}[3pt]J_0(k\rho)k\mathrm{d}k \\
&= \frac{Q}{4\pi\epsilon\sqrt{\rho^2+z^2}} - \frac{Q}{8\pi\epsilon\sqrt{\rho^2+z^2}} - \frac{Q}{8\pi\epsilon\sqrt{\rho^2+z^2}} \\
&= 0
\end{aligned}
$$

In the first case when $\ell \to \infty$, we move the PECs out of sight leaving the solitary point charge at origin; while the case $\ell \to 0$ kills (grounds) and the point charge making potential to zero value! □

Example 5.22. The planes $z = 0$ and $z = h$ are grounded PECs while a point charge Q lies at $(0, 0, a)$ between the planes, that is, $0 < a < h$. See Figure 5.12. Find the electrostatic potential $\phi(\rho, z)$ for $0 < z < h$ using Hankel transforms. After obtaining the answer, allow $h \to \infty$ and move the top plate out of sight retaining the bottom plate as is, and show that the expression tallies with what we get using image method.

Solution: We can write the potential function

$$\phi(\rho, z) = \vartheta(\rho, z) + \frac{Q}{4\pi\epsilon\sqrt{\rho^2+(z-a)^2}}$$

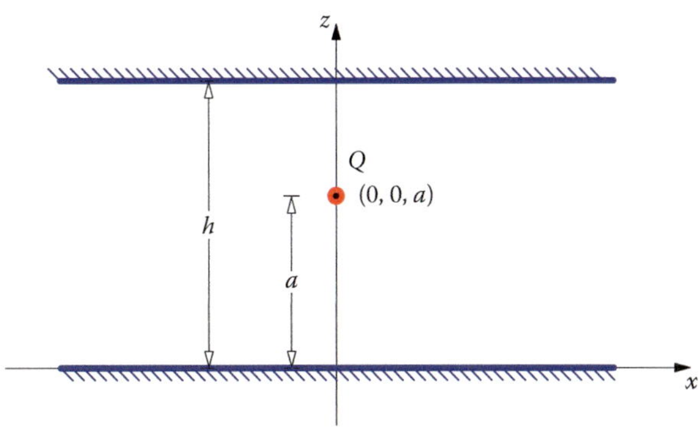

FIGURE 5.12
A point charge Q located at $(0, 0, a)$ between perfectly conducting parallel planes $z = 0$ and $z = h$.

where $\vartheta(\rho, z)$ satisfies the Laplace equation. Let us denote the Hankel transform $\mathcal{H}_0\{\vartheta(\rho, z)\} = \Theta(k, z)$. Then the Laplace's equation in terms of $\Theta(k, z)$ satisfies

$$-k^2\Theta(k,z) + \frac{\mathrm{d}^2\Theta}{\mathrm{d}z^2} = 0$$

whose solution is of the form

$$\Theta(k, z) = A(k)\mathrm{e}^{-kz} + B(k)\mathrm{e}^{kz}$$

where $A(k)$ and $B(k)$ are arbitrary and must be determined in keeping with the boundary conditions

$$\begin{aligned}
\phi(\rho, 0) = 0 \quad &\Rightarrow \quad A(k) + B(k) + \mathcal{H}_0\left\{\frac{Q}{4\pi\epsilon\sqrt{\rho^2 + a^2}}\right\} = 0 \\
&\Rightarrow \quad A(k) + B(k) + \frac{Q\mathrm{e}^{-ka}}{4\pi\epsilon k} = 0 \\
\phi(\rho, h) = 0 \quad &\Rightarrow \quad A(k)\mathrm{e}^{-kh} + B(k)\mathrm{e}^{kh} + \mathcal{H}_0\left\{\frac{Q}{4\pi\epsilon\sqrt{\rho^2 + (h-a)^2}}\right\} = 0 \\
&\Rightarrow \quad A(k) + B(k) + \frac{Q\mathrm{e}^{-k(h-a)}}{4\pi\epsilon k} = 0
\end{aligned}$$

Solving for $A(k)$ and $B(k)$ and utilizing them in potential expression, we get

$$\begin{aligned}
A(k) &= -\frac{Q}{4\pi\epsilon k}\frac{\sinh k(h-a)}{\sinh kh} \\
B(k) &= -\frac{Q}{4\pi\epsilon k}\frac{\mathrm{e}^{-kh}\sinh ka}{\sinh kh} \\
\Theta(k, z) &= -\frac{Q}{4\pi\epsilon k}\left\{\frac{\sinh k(h-a)}{\sinh kh}\mathrm{e}^{-kz} + \frac{\mathrm{e}^{-kh}\sinh ka}{\sinh kh}\mathrm{e}^{kz}\right\} \\
\phi(\rho, z) &= \frac{Q}{4\pi\epsilon\sqrt{\rho^2 + (z-a)^2}} + \int_0^\infty \Theta(k, z)J_0(k\rho)k\mathrm{d}k
\end{aligned}$$

Now let us see what happens if we take the limit $h \to \infty$. First of all,

$$\begin{aligned}
\Theta(k, z)|_{h\to\infty} &= -\frac{Q}{4\pi\epsilon k}\left\{\frac{\sinh k(h-a)}{\sinh kh}\right\}\Bigg|_{h\to\infty}\mathrm{e}^{-kz} \\
&= -\frac{Q}{4\pi\epsilon k}\{\cosh ka - \coth kh \sinh ka\}|_{h\to\infty}\,\mathrm{e}^{-kz} \\
&= -\frac{Q}{4\pi\epsilon k}\{\cosh ka - \sinh ka\}\,\mathrm{e}^{-kz} \\
&= -\frac{Q}{4\pi\epsilon k}\{\exp -k(a+z)\}
\end{aligned}$$

$$\phi(\rho, z)|_{h\to\infty} = \frac{Q}{4\pi\epsilon\sqrt{\rho^2+(z-a)^2}} - \frac{Q}{4\pi\epsilon}\int_0^\infty \left\{\frac{\exp -k(a+z)}{k}\right\} J_0(k\rho)k\mathrm{d}k$$
$$= \frac{Q}{4\pi\epsilon\sqrt{\rho^2+(z-a)^2}} - \frac{Q}{4\pi\epsilon\sqrt{\rho^2+(z+a)^2}}$$

This is what we expect to get using the method of images! □

Example 5.23. Let the dielectric permittivity for the half space $z > 0$ be ϵ_1 and for the half space $z < 0$ be ϵ_2. Consider a point charge Q at $(0, 0, a)$. Find expressions for the electrostatic potential $\phi(\rho, z)$, and the electric field intensity **E**. Taking numerical values $a = 1$, $\epsilon_1 = 2\epsilon_0$, $\epsilon_2 = 8\epsilon_0$, $Q = 4\pi\epsilon_0$ obtain a sketch of equipotential lines and the flux lines.

Solution: Let the potential $\phi(\rho, z)$ be expressed as

$$\phi(\rho, z) = \begin{cases} \vartheta_1(\rho, z) + \frac{Q}{4\pi\epsilon_1\sqrt{\rho^2+(z-a)^2}} & z > 0 \\ \vartheta_2(\rho, z) & z < 0 \end{cases}$$

where $\vartheta_1(\rho, z)$ and $\vartheta_2(\rho, z)$ satisfy the Laplace's equation. Applying Hankel transform of order zero from the variable ρ to k and denoting $\mathcal{H}_0\{\vartheta(\rho, z)\} = \Theta(k, z)$, we find that

$$\Theta_1(k, z) = A(k)\mathrm{e}^{-kz}$$
$$\Theta_2(k, z) = B(k)\mathrm{e}^{kz}$$

The reason for this choice is based on the fact that potential should tend to zero for $z \to \pm\infty$. Continuity in potential and the normal component of electric flux density yield

$$\phi(\rho, 0^+) = \phi(\rho, 0^-) \Rightarrow A(k) + \frac{Q}{4\pi\epsilon_1}\frac{\mathrm{e}^{-ka}}{k} = B(k)$$

$$\epsilon_1 \frac{\phi(\rho, z)}{\mathrm{d}z}\bigg|_{z=0^+} = \epsilon_2 \frac{\phi(\rho, z)}{\mathrm{d}z}\bigg|_{z=0^-} \Rightarrow -\epsilon_1 kA(k) + \epsilon_1 k\frac{Q}{4\pi\epsilon_1}\frac{\mathrm{e}^{-ka}}{k} = \epsilon_2 kB(k)$$

Solving for $A(k)$ and $B(k)$, we can get

$$\Theta_1(k, z) = \frac{\epsilon_1 - \epsilon_2}{\epsilon_2 + \epsilon_1}\frac{Q}{4\pi\epsilon_1}\frac{\mathrm{e}^{-ka}}{k}\mathrm{e}^{-kz}$$
$$\Theta_2(k, z) = \frac{2\epsilon_1}{\epsilon_2 + \epsilon_1}\frac{Q}{4\pi\epsilon_1}\frac{\mathrm{e}^{-ka}}{k}\mathrm{e}^{kz}$$

Inverting we get

$$\phi_1(\rho, z) = \frac{\epsilon_1 - \epsilon_2}{\epsilon_2 + \epsilon_1}\frac{Q}{4\pi\epsilon_1}\frac{1}{\sqrt{\rho^2+(z+a)^2}} + \frac{Q}{4\pi\epsilon_1}\frac{1}{\sqrt{\rho^2+(z-a)^2}}$$
$$\phi_2(\rho, z) = \frac{Q}{2\pi(\epsilon_2 + \epsilon_1)}\frac{1}{\sqrt{\rho^2+(z-a)^2}}$$

Once again, a quick check can be made to satisfy ourselves by considering trivial cases. Suppose $\epsilon_1 = \epsilon_2$, it means there is only one medium. Then the potential becomes

$$\begin{aligned}
\phi_1(\rho, z) &= 0 + \frac{Q}{4\pi\epsilon_1}\frac{1}{\sqrt{\rho^2 + (z-a)^2}} \\
\phi_2(\rho, z) &= \frac{Q}{2\pi(\epsilon_1 + \epsilon_1)}\frac{1}{\sqrt{\rho^2 + (z-a)^2}} \\
\phi_1(\rho, z) = \phi_2(\rho, z) &= \frac{Q}{4\pi\epsilon_1}\frac{1}{\sqrt{\rho^2 + (z-a)^2}}
\end{aligned}$$

which is good. Likewise, the fact that $E = \frac{D}{\epsilon}$ kills the electric field when $\epsilon \to \infty$. So if we let $\epsilon_2 \to \infty$ it has the effect of bringing a PEC along the xy-plane. The potential becomes

$$\begin{aligned}
\phi_1(\rho, z) &= -\frac{Q}{4\pi\epsilon_1}\frac{1}{\sqrt{\rho^2 + (z+a)^2}} + \frac{Q}{4\pi\epsilon_1}\frac{1}{\sqrt{\rho^2 + (z-a)^2}} \\
\phi_2(\rho, z) &= 0
\end{aligned}$$

This situation simulates the problem of a point charge above a PEC which is solved using image method. We see that happening. Next, using gradient we can express the electric field intensity as

$$\begin{aligned}
\mathbf{E}_1 &= -\nabla\phi_1(\rho, z) \\
\mathbf{E}_2 &= -\nabla\phi_2(\rho, z)
\end{aligned}$$

Last, substituting numerical values, we can get the plots shown in Figure 5.13. □

Example 5.24. An infinite cylindrical shell of radius a carries a time harmonic current in the z-direction on its surface with a density J_s. Find the radiated fields. The shell is shown in Figure 5.14.

Solution: For the given distribution of current source, symmetry reveals that $\mathbf{A} = A_z(\rho)\mathbf{a_z}$. We shall express the Helmholtz equation for $\mathbf{A}$ in cylindrical coordinates:

$$\frac{1}{\rho}\frac{\mathrm{d}}{\mathrm{d}\rho}\left(\rho\frac{\mathrm{d}A_z}{\mathrm{d}\rho}\right) + k^2 A_z = -\frac{\mu J_s \delta(\rho - a)}{2\pi\rho}$$

Let us take Hankel transform on both sides. The Hankel transform of $A_z(\rho, k)$ is denoted by $\tilde{A}_z(\xi, k)$.

$$\begin{aligned}
\left(-\xi^2 + k^2\right)\tilde{A}_z(\xi, k) &= -\frac{\mu J_s}{2\pi}\mathrm{J}_0(\xi a) \\
\tilde{A}_z(\xi, k) &= \frac{\mu J_s}{2\pi}\frac{\mathrm{J}_0(\xi a)}{(\xi^2 - k^2)}
\end{aligned}$$

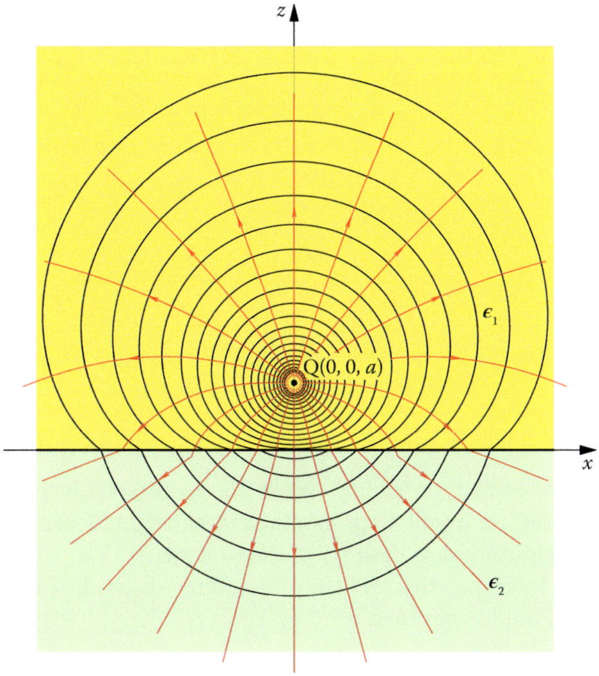

FIGURE 5.13
Equipotential and flux lines in case of a point charge Q located at $(0, 0, a)$, with $a = 1$. The permittivity for $z > 0$ is $\epsilon_1 = 2\epsilon_0$ and for $z < 0$ is $\epsilon_2 = 8\epsilon_0$. Observe the refraction of flux lines and equipotential lines.

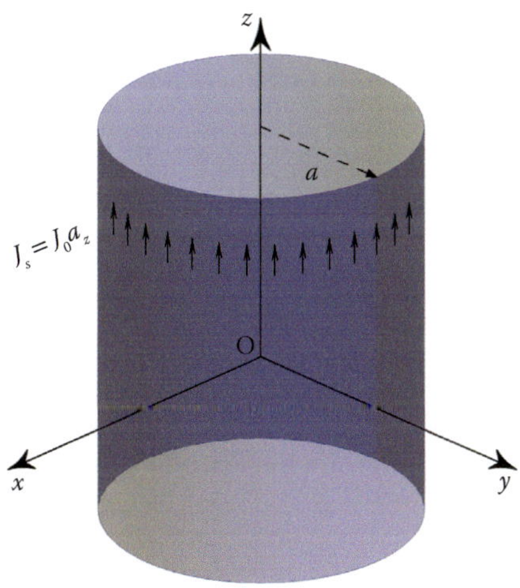

FIGURE 5.14
An infinite cylindrical sheet of radius a around the z-axis carrying uniform surface current.

Finding inverse Hankel transform,

$$\begin{aligned} A_z(\rho, k) &= \frac{\mu J_s}{2\pi}\int_0^{\infty}\frac{J_0(\xi a)J_0(\xi\rho)}{(\xi^2-k^2)}\xi d\xi \\ &= \frac{\mu J_s}{4\pi}\int_0^{\infty}\frac{J_0(\xi\rho)}{(\xi^2-k^2)}\left(H_0^{(1)}(\xi a)+H_0^{(2)}(\xi a)\right)\xi d\xi \\ &= \frac{\mu J_s}{4\pi}\int_0^{\infty}\frac{J_0(\xi\rho)}{(\xi^2-k^2)}\left(H_0^{(1)}(\xi a)-H_0^{(1)}(-\xi a)\right)\xi d\xi \\ &= \frac{\mu J_s}{4\pi}\int_{-\infty}^{\infty}\frac{J_0(\xi\rho)H_0^{(1)}(\xi a)}{(\xi^2-k^2)}\xi d\xi \end{aligned}$$

The above integral can be solved using contour integration. Let us denote the integrand as $f(\xi) = \frac{J_0(\xi\rho)H_0^{(1)}(\xi a)\,\xi}{(\xi^2-k^2)}$. Since $H_0^{(1)}(\xi a)$ has a logarithmic branch point at the origin, we might impose a branch cut along $\text{Re}\{\xi\} < 0$. We shall consider two separate cases: case (i) $\rho < a$ and case (ii) $\rho > a$. For handling these two cases there are two choices of contours shown in Figure 5.15a and b, respectively. For the case $\rho < a$

$$\int_{\Gamma_1} f(\xi)d\xi + \int_{\Gamma_2} f(\xi)d\xi + \int_{\Gamma_3} f(\xi)d\xi + \int_{\Gamma_4} f(\xi)d\xi = 2\pi j\left\{\operatorname*{Res}_{\xi=k} f(\xi)\right\}$$

If we allow $R \to \infty$ and $r \to 0$, the contributions of the integral along Γ_2 vanishes, and so does Γ_4. To gain some insight into the latter, it is apt to recall the large argument formulas for the Bessel and Hankel functions:

$$\begin{aligned} J_0(\xi\rho) &\approx \sqrt{\frac{1}{2\pi\xi\rho}}\left\{\exp\left[j\left(\xi\rho-\frac{\pi}{4}\right)\right]+\exp\left[-j\left(\xi\rho-\frac{\pi}{4}\right)\right]\right\} \\ H_0^{(1)}(\xi a) &\approx \sqrt{\frac{2}{\pi\xi a}}\exp\left[j\left(\xi a-\frac{\pi}{4}\right)\right] \\ J_0(\xi\rho)H_0^{(1)}(\xi\rho a) &\approx \text{a term with} \exp\left[j\left(\xi(a+\rho)\right)\right] \\ &+ \text{another term with} \exp\left[j\left(\xi(a-\rho)\right)\right] \end{aligned}$$

For the case $0 < \rho < a$, both $(a+\rho) > 0$ and $(a-\rho) > 0$, making integral tend to zero in the light of Jordan's lemma. The integral along Γ_1 and Γ_3 put together is the sought for integral which can be evaluated from the residue at $\xi = k$. Thus,

$$\begin{aligned} A_z(\rho, k) &= \frac{\mu J_s}{4\pi}\times 2\pi j \times \left.\frac{J_0(\xi\rho)H_0^{(1)}(\xi a)}{(\xi+k)}\xi\right|_{\xi=k} \\ &= \frac{j\mu J_s}{4}J_0(k\rho)H_0^{(1)}(ka) \end{aligned}$$

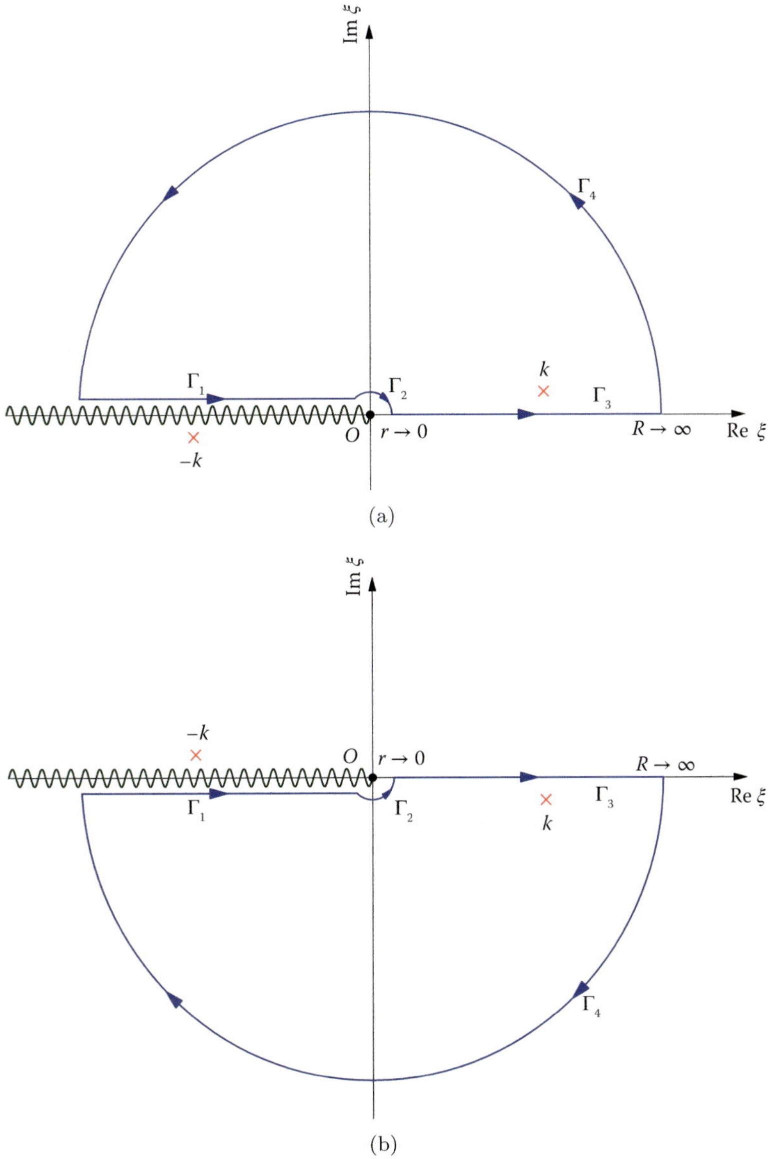

FIGURE 5.15
(a) Contour for the case $\rho < a$. (b) Contour for the case $\rho > a$.

A similar treatment for the case $\rho > a$ can be developed using the contour shown in Figure 5.15b . Combining both results we get

$$A_z(\rho, k) \quad = \quad \begin{cases} \frac{j\mu J_s}{4} \mathrm{J}_0(k\rho)\mathrm{H}_0^{(1)}(ka) & \text{for } \rho < a \\ \frac{j\mu J_s}{4} \mathrm{J}_0(ka)\mathrm{H}_0^{(1)}(k\rho) & \text{for } \rho > a \end{cases}$$

□

5.4 The Mellin Transform

Toward defining the Mellin transform pair, let us start with the inversion of the Fourier transform. Suppose $f(x)$ and $F(k)$ constitute a Fourier transform pair. Then, we can functionally write

$$f(x) = \frac{1}{2\pi}\int_{\omega=-\infty}^{\infty} e^{-j\omega x}\int_{u=-\infty}^{\infty} f(u)e^{j\omega u}\,du\,d\omega$$

Let us introduce the dummy variables $r = e^{x}$ and $\rho = e^{u}$ and also let $k = \sigma + j\omega$, where σ is a fixed constant. So we find that

$$\begin{aligned} f(\log r) &= \frac{1}{2\pi j}\int_{k=\sigma-j\infty}^{\sigma+j\infty} r^{-j\omega}\int_{\rho=0}^{\infty} f(\log\rho)\rho^{j\omega-1}\,d\rho\,dk \\ &= \frac{1}{2\pi j}\int_{k=\sigma-j\infty}^{\sigma+j\infty} r^{\sigma-k}\int_{\rho=0}^{\infty} f(\log\rho)\rho^{k-\sigma-1}\,d\rho\,dk \\ f(\log r)r^{-\sigma} &= \frac{1}{2\pi j}\int_{k=\sigma-j\infty}^{\sigma+j\infty} r^{-k}\int_{\rho=0}^{\infty} f(\log\rho)\rho^{-\sigma}\rho^{k-1}\,d\rho\,dk \end{aligned}$$

Now let us define $V(\rho) = f(\log\rho)\rho^{-\sigma}$. We get

$$V(r) = \frac{1}{2\pi j}\int_{k=\sigma-j\infty}^{\sigma+j\infty} r^{-k}V_M(k)\,dk$$

where

$$V_M(k) = \int_{\rho=0}^{\infty} V(\rho)\rho^{k-1}\,d\rho$$

We thus arrive at the definition of the Mellin transform and its inverse as

$$\begin{aligned} \mathcal{M}\{V(\rho)\} = V_M(k) &= \int_{\rho=0}^{\infty} V(\rho)\rho^{k-1}\,d\rho \\ V(\rho) &= \frac{1}{2\pi j}\int_{k=\sigma-j\infty}^{\sigma+j\infty} r^{-k}V_M(k)\,dk \end{aligned}$$

Next, let us try to evaluate the Mellin transform of a few simple functions.

Example 5.25. Find the Mellin transform of

(a) $e^{-\rho}$
(b) $\cos\rho$
(c) $\sin\rho$
(d) $\frac{1}{1+\rho}$

Solution: Using the definition,

(a)
$$\begin{aligned}\mathcal{M}\left\{e^{-\rho}\right\} &= \int_{-\infty}^{\infty} \rho^{k-1} e^{-\rho} d\rho \\ &= \Gamma(k) \qquad\qquad \mathfrak{Re}(k) > 0\end{aligned}$$

(b)
$$\begin{aligned}\mathcal{M}\left\{\cos\rho\right\} &= \int_{-\infty}^{\infty} \rho^{k-1} \cos\rho d\rho \\ &= \Gamma(k)\cos\left(\frac{1}{2}\pi k\right) \qquad\qquad 0 < \mathfrak{Re}(k) < 1\end{aligned}$$

(c)
$$\begin{aligned}\mathcal{M}\left\{\sin\rho\right\} &= \int_{-\infty}^{\infty} \rho^{k-1} \sin\rho d\rho \\ &= \Gamma(k)\sin\left(\frac{1}{2}\pi k\right) \qquad\qquad 0 < \mathfrak{Re}(k) < 1\end{aligned}$$

(d)
$$\begin{aligned}\mathcal{M}\left\{\frac{1}{1+\rho}\right\} &= \int_{-\infty}^{\infty} \rho^{k-1} \frac{1}{1+\rho} d\rho \\ &= \Gamma(k)\Gamma(1-k) \qquad\qquad 0 < \mathfrak{Re}(k) < 1 \\ &= \frac{\pi}{\sin\pi k} \qquad\qquad 0 < \mathfrak{Re}(k) < 1\end{aligned}$$

□

5.4.1 Properties of the Mellin Transform

Linearity: The Mellin transform obeys linearity in that if A and B are arbitrary constants, the Mellin transform of $Af(\rho) + Bg(\rho)$ is given by

$$\mathcal{M}\left\{Af(\rho) + Bg(\rho)\right\} = A\mathcal{M}\left\{f(\rho)\right\} + B\mathcal{M}\left\{g(\rho)\right\}$$

Scaling: The Mellin transform of $f(a\rho)$, where $a > 0$, is given by

$$\mathcal{M}\left\{f(a\rho)\right\} = \frac{1}{a^k}\mathcal{M}\left\{f(\rho)\right\}$$

Translation: If the Mellin transform of $f(\rho)$ is $f_M(k)$, then the Mellin transform of $\rho^a f(\rho)$ is given by

$$\mathcal{M}\left\{\rho^a f(\rho)\right\} = f_M(k+a)$$

Exponentiation: If the Mellin transform of $f(\rho)$ is $f_M(k)$, then the Mellin transform of $f(\rho^a)$, where $a > 0$, is given by

$$\mathcal{M}\left\{f(\rho^a)\right\} = \frac{1}{a} f_M\left(\frac{k}{a}\right)$$

Differentiation: If $f(\rho)$ is continuous on $\rho \geq 0$ and has the Mellin transform of $f_M(k)$, then the Mellin transform of $f'(\rho^a)$ is given by

$$\begin{aligned}\mathcal{M}\{f'(\rho)\} &= \int_{-\infty}^{\infty} \rho^{k-1} f'(\rho)\mathrm{d}\rho \\ &= \rho^{k-1} f(\rho)\big|_0^{\infty} - (k-1)\int_{-\infty}^{\infty} \rho^{k-2} f(\rho)\mathrm{d}\rho\end{aligned}$$

Now if there exist two bounds σ_1 and σ_2 such that the first term vanishes at both the limits 0 and ∞, then we can say

$$\mathcal{M}\{f'(\rho)\} = -(k-1)f_M(k-1)$$

Similarly,

$$\mathcal{M}\{f''(\rho)\} = (k-1)(k-2)f_M(k-2)$$

Eigen property: The Mellin transform of the operator $\left(\rho^2 \frac{\mathrm{d}^2}{\mathrm{d}\rho^2} + \rho \frac{\mathrm{d}}{\mathrm{d}\rho}\right)$ is k^2. Hence,

$$\mathcal{M}\left\{\rho^2 \frac{\mathrm{d}^2 f}{\mathrm{d}\rho^2} + \rho \frac{\mathrm{d}f}{\mathrm{d}\rho}\right\} = k^2 f_M(k)$$

This property can be easily proved by using the differentiation and translation properties. Next, we provide two examples showing the use of Mellin transform in solving problems involving wedges.

Example 5.26. In the cylindrical coordinate system, consider three conducting (PEC) boundaries: the semi-infinite plane $\varphi = 0$, the infinite strip of finite width described by $\varphi = \alpha, 0 < \rho < a$ and the semi-infinite plane $\varphi = \alpha, \rho > a$. They are held at 0, V_0, and 0 volts, respectively. To have separation between them, assume infinitesimal insulating gaps. Solve the Laplace equation in the region $0 < \varphi < \alpha$ and express the potential function $V(\rho, \varphi)$. Use the Mellin transform. Plot the equipotential lines and flux lines.

Solution: Let the Mellin transform $\mathcal{M}\{V(\rho, \varphi)\}$ be denoted by $V_M(k, \varphi)$. In terms of the transform, the Laplace's equation becomes

$$k^2 V_M + \frac{\mathrm{d}^2 V_M}{\mathrm{d}\varphi^2} = 0$$

whose general solution is of the form

$$V_M(k, \varphi) = A \sin k\varphi + B \cos k\varphi$$

To satisfy the condition that $V(\rho, 0) = 0$, we need to make $B = 0$ and write

$$V_M(k, \varphi) = A \sin k\varphi$$

But for $\varphi = \alpha$,

$$V(\rho, \alpha) = \begin{cases} V_0 & \text{for} \quad 0 < \rho < a \\ 0 & \text{for} \quad \rho > a \end{cases}$$

In terms of $V_M(k, \varphi)$, we can write this as

$$V_M(k, \alpha) = V_0 \frac{a^k}{k}$$

Solving for A, we can write

$$V_M(k, \varphi) = V_0 \frac{a^k \sin k\varphi}{k \sin k\alpha}$$

The inversion integral can be set down as

$$V(\rho, \varphi) = \frac{V_0}{2\pi j} \int_{\sigma - j\infty}^{\sigma + j\infty} \frac{\sin k\varphi}{k \sin k\alpha} \left(\frac{a}{\rho}\right)^k \mathrm{d}k$$

Using calculus of residues, we can evaluate the integral and write

$$V(\rho, \varphi) = \begin{cases} V_0 \frac{\varphi}{\alpha} + \frac{V_0}{\pi} \sum_{n=1}^{\infty} \frac{(-1)^{n+1}}{n} \left(\frac{\rho}{a}\right)^{n\pi/\alpha} \sin\left(\frac{n\pi\varphi}{\alpha}\right) & 0 < \rho < a \\ \\ \frac{V_0}{\pi} \sum_{n=1}^{\infty} \frac{(-1)^{n+1}}{n} \left(\frac{a}{\rho}\right)^{n\pi/\alpha} \sin\left(\frac{n\pi\varphi}{\alpha}\right) & a < \rho < \infty \end{cases}$$

It is convenient to obtain a closed form expression for an infinite series. Although it may not be very obvious how to do this in a general case, particularly for the present, we can succeed as follows: Recall that whenever $|z| < 1$, the function $\log(1 + z)$ has the following Taylor series expansion:

$$\log(1 + z) = \sum_{n=1}^{\infty} (-1)^{n+1} \frac{z^n}{n}$$

Next, let $z = |z| \mathrm{e}^{j\angle z}$ in polar representation. Then

$$\begin{aligned} \log(1 + z) &= \log\{\sqrt{1 + |z|^2 + 2|z| \cos \angle z}\} + j \arctan\left\{\frac{|z| \sin \angle z}{1 + |z| \cos \angle z}\right\} \\ &= \sum_{n=1}^{\infty} (-1)^{n+1} \frac{|z|^n}{n} \mathrm{e}^{jn\angle z} \end{aligned}$$

Taking the imaginary parts both sides,

$$\sum_{n=1}^{\infty} (-1)^{n+1} \frac{|z|^n}{n} \sin n\angle z = \arctan\left\{\frac{|z| \sin \angle z}{1 + |z| \cos \angle z}\right\}$$

Let us choose

$$z = \begin{cases} \left(\frac{\rho}{a}\right)^{\frac{\pi}{\alpha}} e^{\frac{j\pi\varphi}{\alpha}} & 0 < \rho < a \\ \left(\frac{a}{\rho}\right)^{\frac{\pi}{\alpha}} e^{\frac{j\pi\varphi}{\alpha}} & a < \rho < \infty \end{cases}$$

Then

$$V(\rho,\varphi) = \begin{cases} V_0\frac{\varphi}{\alpha} + \frac{V_0}{\pi}\arctan\left\{\frac{\left(\frac{\rho}{a}\right)^{\frac{\pi}{\alpha}}\sin\frac{\pi\varphi}{\alpha}}{1+\left(\frac{\rho}{a}\right)^{\frac{\pi}{\alpha}}\cos\frac{\pi\varphi}{\alpha}}\right\} & 0 < \rho < a \\ \frac{V_0}{\pi}\arctan\left\{\frac{\left(\frac{a}{\rho}\right)^{\frac{\pi}{\alpha}}\sin\frac{\pi\varphi}{\alpha}}{1+\left(\frac{a}{\rho}\right)^{\frac{\pi}{\alpha}}\cos\frac{\pi\varphi}{\alpha}}\right\} & a < \rho < \infty \end{cases}$$

The equipotential lines and the flux lines in the solution regions are shown in Figure 5.16.

□

Example 5.27. Consider an infinite uniform line charge of density $\rho_l = \rho_0$ C/m lying along $\rho = a$, $\phi = 0$. Let the semi-infinite planes $\varphi = \pm\alpha$ be PECs and let them be at zero potential. Find an expression for the potential in the region $|\varphi| < \alpha$.

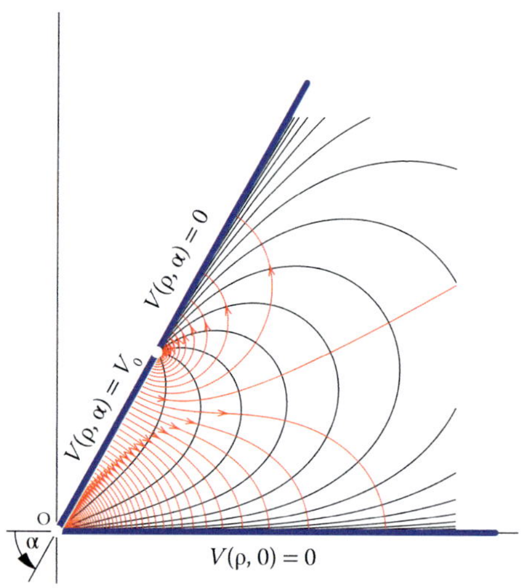

FIGURE 5.16
Equipotential and flux lines obtained using Mellin transform.

Solution: The potential function in this satisfies:

$$\begin{aligned}
\frac{1}{\rho}\frac{\partial}{\partial\rho}\left(\rho\frac{\partial V}{\partial\rho}\right) + \frac{1}{\rho^2}\frac{\partial^2 V}{\partial\varphi^2} &= -\frac{\rho_0}{\epsilon\rho}\delta(\rho - a)\delta(\varphi) \\
\Rightarrow \quad \rho^2\frac{\partial^2 V}{\partial\rho^2} + \rho\frac{\partial V}{\partial\rho} + \frac{\partial^2 V}{\partial\varphi^2} &= -\frac{\rho_0}{\epsilon}\rho\delta(\rho - a)\delta(\varphi)
\end{aligned}$$

Let the Mellin transform $\mathcal{M}\{V(\rho,\varphi)\}$ be denoted by $V_M(k,\varphi)$. In terms of the transform, the Poisson's equation becomes

$$\begin{aligned}
(k + k^2)V_M + (-pV_M) + \frac{\mathrm{d}^2 V_M}{\mathrm{d}\varphi^2} &= -\frac{\rho_0}{\epsilon}a^k\delta(\varphi) \\
k^2 V_M + \frac{\mathrm{d}^2 V_M}{\mathrm{d}\varphi^2} &= -\frac{\rho_0}{\epsilon}a^k\delta(\varphi)
\end{aligned}$$

In order that

$$\begin{aligned}
V_M(\rho, \pm\alpha) &= 0 \\
V_M(\rho, +\varphi) &= V_M(\rho, -\varphi) \\
\left.\frac{\mathrm{d}V_M}{\mathrm{d}\varphi}\right|_{\varphi=0^-}^{\varphi^+} &= -\frac{\rho_0}{\epsilon}a^k
\end{aligned}$$

a possible solution is

$$V_M(k,\varphi) = \frac{\rho_0}{2\epsilon}\frac{\sin k(\alpha - |\varphi|)}{k\cos k\alpha}a^k$$

Inverting, we get the expression for the potential

$$\begin{aligned}
V(\rho,\varphi) &= \frac{\rho_0}{4\pi\epsilon j}\int_\Gamma \frac{\sin k(\alpha - |\varphi|)}{k\cos k\alpha}\left(\frac{a}{\rho}\right)^k \mathrm{d}k \\
&= \frac{\rho_0}{2\epsilon}\sum_{m \text{ odd}} \left.\frac{\sin(k\alpha - k|\varphi|)}{(-k\alpha)\sin k\alpha}\left(\frac{a}{\rho}\right)^k\right|_{k\alpha=-m\pi/2} \\
&= \frac{\rho_0}{2\epsilon}\sum_{m \text{ odd}} \left.\frac{\sin k\alpha\cos k\varphi - \cos k\alpha\sin k|\varphi|}{(-k\alpha)\sin k\alpha}\left(\frac{a}{\rho}\right)^k\right|_{k\alpha=-m\pi/2} \\
&= \frac{\rho_0}{2\epsilon}\sum_{m \text{ odd}} \left.\frac{\cos k\varphi - \cot k\alpha\sin k|\varphi|}{(-k\alpha)}\left(\frac{a}{\rho}\right)^k\right|_{k=-m\pi/(2\alpha)} \\
&= \frac{\rho_0}{\pi\epsilon}\sum_{m \text{ odd}} \frac{\cos\left(\frac{m\pi}{2\alpha}\varphi\right)}{m}\left(\frac{\rho}{a}\right)^{\frac{m\pi}{2\alpha}} \\
&= \frac{\rho_0}{4\pi\epsilon}\log\left\{\frac{\rho^{\pi/\alpha} + 2(a\rho)^{\pi/(2\alpha)}\cos(\pi\varphi/(2\alpha)) + a^{\pi/\alpha}}{\rho^{\pi/\alpha} - 2(a\rho)^{\pi/(2\alpha)}\cos(\pi\varphi/(2\alpha)) + a^{\pi/\alpha}}\right\}
\end{aligned}$$

As a special case if $\alpha = \pi/2$, the wedge occupies the whole yz-plane and the potential becomes

$$V(\rho, \varphi) = \frac{\rho_0}{4\pi\epsilon} \log \left\{ \frac{\rho^2 + 2a\rho\cos\varphi + a^2}{\rho^2 - 2a\rho\cos\varphi + a^2} \right\}$$

which is the expected answer obtained from method of images. For the special case of $\alpha = \pi/4$, with $a = 8$, we show the equipotential lines and the flux lines in Figure 5.17. □

5.5 Laplace Transform

The Fourier transform discussed at the start of the present chapter has some limitations. Many a function of time or position commonly occurring in engineering and science do not possess the Fourier transform for want of absolute integrability. Such functions can be integrated by multiplying with a decaying

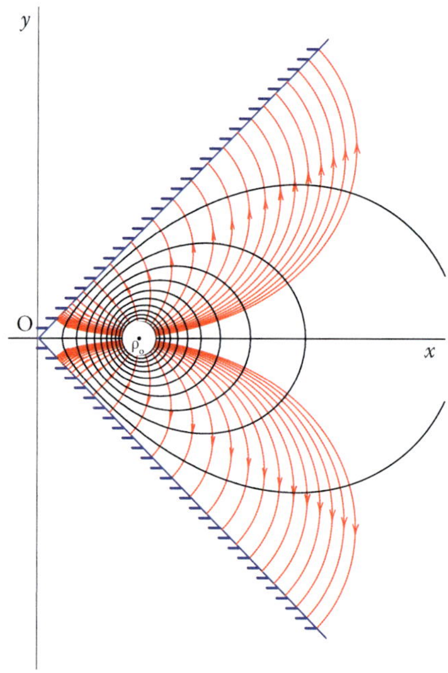

FIGURE 5.17
An infinite line charge ρ_0 located parallel to the z-axis and passing through $(a, 0)$ flanked by grounded PECs along $\varphi = \pm\pi/4$. For the choice $a = 8$, the equipotential and flux lines are depicted.

exponential, which is the key aspect in the development of the Laplace transform. The two variables x and k in the definition of the Fourier transform could be time and angular frequency. Or they could be space and spacial frequency (or the wave number). To interpret the variables occurring in the Laplace transform, we need to generalize the notion of frequency as a complex quantity. A more general version of the two-sided Fourier transform is the two-sided (bilateral) Laplace transform. Although the Laplace transform could be defined for two-sided functions defined over the domain of the entire real line, it is enough to consider only the one-sided or causal functions that might be considered to be zero for the negative values in the function domain. The resulting transform is called the one-sided or unilateral Laplace transform. Unless otherwise stated, we shall only deal with the one-sided Laplace transform. The Laplace transform of $f(t)$ denoted by $\mathcal{L}\{f(t)\} = F(s)$, where $s \in \mathbb{C}$ is given by

$$F(s) = \int_0^\infty f(t)\mathrm{e}^{-st}\mathrm{d}t \tag{5.36}$$

and the inverse Laplace transform is found by evaluating the inversion integral through

$$f(t) = \frac{1}{2\pi j}\int_{\sigma - j\infty}^{\sigma + j\infty} F(s)\mathrm{e}^{st}\mathrm{d}s \tag{5.37}$$

In order for the integral in Equation 5.37 to converge, it is required that the variable $s = \sigma + j\omega$ must lie in an open subset of $\mathbb{C}$ called the *region of convergence* or ROC. The path along which the integral in Equation 5.37 is evaluated is to lie in ROC with the end points reaching from $-j\infty$ to $j\infty$. One might classify functions that are Laplace transformable into four groups based on their support as (i) left-sided, (ii) right-sided, (iii) two-sided, and (iv) finite-support functions. The ROCs of these functions take on the forms: (i) $\sigma < \sigma_2$, (ii) $\sigma > \sigma_1$, (iii) $\sigma_1 < \sigma < \sigma_2$, and (iv) $-\infty < \sigma < \infty$. For most of the simpler functions the Laplace transform is a rational function in s. The roots of the numerator are called zeros and the roots of the denominator are called poles. No pole is an interior point in ROC. But a pole can be a limit point.

Example 5.28. Find the Laplace transforms of the following functions. Note that the Heaviside step function is denoted by $u(t)$.

(a) $u(t)$
(b) $\mathrm{e}^{at}u(t)$
(c) $\sin at\, u(t)$
(d) $\cos at\, u(t)$

Solution: By direct integration, we can find the transforms in a straightforward manner.

(a)
$$\begin{aligned}\mathcal{L}\{u(t)\} &= \int_0^{\infty} (1)\mathrm{e}^{-st}\mathrm{d}t \\ &= \left.\frac{\mathrm{e}^{-st}}{-s}\right|_0^{\infty} \\ &= \frac{1}{s}\end{aligned}$$

(b)
$$\begin{aligned}\mathcal{L}\{\mathrm{e}^{at}u(t)\} &= \int_0^{\infty} \mathrm{e}^{at-st}\mathrm{d}t \\ &= \left.\frac{\mathrm{e}^{-(s-a)t}}{-(s-a)}\right|_0^{\infty} \\ &= \frac{1}{s-a}\end{aligned}$$

(c)
$$\begin{aligned}\mathcal{L}\{\sin at\, u(t)\} &= \int_0^{\infty} \sin at\, \mathrm{e}^{-st}\mathrm{d}t \\ &= \left.\frac{-\mathrm{e}^{-st}}{s^2+a^2}(s\sin at + a\cos at)\right|_0^{\infty} \\ &= \frac{a}{s^2+a^2}\end{aligned}$$

(d)
$$\begin{aligned}\mathcal{L}\{\cos at\, u(t)\} &= \int_0^{\infty} \cos at\, \mathrm{e}^{-st}\mathrm{d}t \\ &= \left.\frac{-\mathrm{e}^{-st}}{s^2+a^2}(s\cos at - a\sin at)\right|_0^{\infty} \\ &= \frac{s}{s^2+a^2}\end{aligned}$$

□

5.5.1 Properties of the Laplace Transform

The Laplace transform has very interesting properties. We shall consider some of them.

Linearity: Based on the linearity of integral operation, we can state

$$\mathcal{L}\{C_1 f(t) + C_2 g(t)\} = C_1\mathcal{L}\{f(t)\} + C_2\mathcal{L}\{g(t)\} \tag{5.38}$$

where C_1 and C_2 are arbitrary constants and the ROC of the linear combination is the intersection of individual ROCs.

Shifting in the function domain: If the function $f(t)$ whose Laplace transform is $F(s)$ is shifted in time t, by change of variables we can easily derive that

$$\mathcal{L}\{f(t-a)\} = \mathrm{e}^{-as}\mathcal{L}\{f(t)\} \tag{5.39}$$

Shifting in the transform domain: If the function $f(t)$ whose Laplace transform is $F(s)$ is multiplied in time by e^{at}, we find that

$$\mathcal{L}\{f(t)e^{at}\} = F(s-a) \tag{5.40}$$

Scaling: Suppose the function $f(t)$ has the Laplace transform $F(s)$, then

$$\mathcal{L}\{f(at)\} = \frac{1}{|a|}F\left(\frac{s}{a}\right) \tag{5.41}$$

Differentiation in the function domain: Suppose the function $f(t)$ has the Laplace transform $F(s)$, then the transform of the derivative is found as

$$\begin{aligned}\mathcal{L}\{f'(t)\} &= \int_0^\infty f'(t)e^{-st}dt &(5.42)\\ &= e^{-st}f(t)\big|_0^\infty + s\int_0^\infty f(t)e^{-st}dt &(5.43)\\ &= sF(s) - f(0) &(5.44)\end{aligned}$$

Now suppose $f(t), f'(t), \ldots, f^{n-1}(t)$ are all continuous on $t \geq 0$ and if $f^n(t)$ is piecewise continuous on $t \geq 0$ and if they are of finite exponential order, the differentiation property can be extended as

$$\mathcal{L}\{f^{(n)}(t)\} = s^nF(s) - s^{n-1}f(0) - s^{n-2}f'(0) - \ldots - f^{n-1}(0) \tag{5.45}$$

Convolution in the function domain: Suppose $f(t)$ and $g(t)$ have their Laplace transforms given by $F(s)$ and $G(s)$, respectively, the convolution of $f(t)$ and $g(t)$ has the Laplace transform given by $F(s)G(s)$. Let us verify.

$$\begin{aligned}\mathcal{L}\{f(t)\star g(t)\} &= \int_0^\infty\int_0^\infty f(\tau)g(t-\tau)d\tau e^{-st}dt &(5.46)\\ &= \int_0^\infty\int_0^\infty f(\tau)g(t-\tau)d\tau e^{-s(t-\tau)}e^{-s\tau}dt &(5.47)\\ &= \int_0^\infty\left[\int_0^\infty g(u)e^{-su}du\right]f(\tau)e^{-s\tau}d\tau &(5.48)\\ &= F(s)G(s) &(5.49)\end{aligned}$$

Here, it may be noted that the limits of integration for convolution have been extended without causing error in the result.

Integration in the function domain: Convolving a given function with the unit step results in the running integral as given by

$$f(t) \star u(t) \quad = \quad \int_0^t f(\tau) d\tau\}$$

Now using the convolution property, the transform of the integral of $f(t)$ that is piecewise continuous on $t \geq 0$ can be readily obtained as

$$\mathcal{L}\left\{\int_0^t f(\tau) d\tau\right\} \quad = \quad \mathcal{L}\left\{f(t) \star u(t)\right\} \tag{5.50}$$

$$= \quad \frac{1}{s}\mathcal{L}\left\{f(t)\right\} \tag{5.51}$$

Example 5.29. Consider a semi-infinite isotropic medium of permittivity ϵ and permeability μ over the domain $z \geq 0$. The x-component of electric field is initially zero for all z. At the boundary $z = 0$ it is made to vary according to the time domain function $\psi(t)$ for $t \geq 0$. Use Laplace transform technique to solve the wave equation:

$$\frac{\partial^2 E_x}{\partial t^2} = \frac{1}{\mu\epsilon} \frac{\partial^2 E_x}{\partial z^2}$$

Solution: Let v denote the phase velocity $\frac{1}{\sqrt{\mu\epsilon}}$. The boundary/initial conditions for this problem are

$$E_x(z, 0) \quad = \quad 0$$

$$\frac{\partial}{\partial t} E_x(z, 0) \quad = \quad 0$$

$$E_x(0, t) \quad = \quad \psi(t)$$

Let us define the Laplace transform from the time variable t to the transform variable s and denote $\mathcal{L}\{E_x(z,t)\} = U(z,s)$. Also let $\mathcal{L}\{\psi(t)\} = \Psi(s)$. Applying the transform to the wave equation,

$$v^2 \frac{\mathrm{d}^2 U}{\mathrm{d}z^2} - s^2 U(z,s) \quad = \quad 0 \quad \text{for } z > 0$$

A general solution for this is of the form

$$U(z,s) \quad = \quad A(s)\mathrm{e}^{-\frac{z}{v}s} + B(s)\mathrm{e}^{\frac{z}{v}s}$$

To have finite value for all $z > 0$, we require that $B(s) = 0$ and since

$$U(0,s) \quad = \quad \Psi(s) = A(p)$$

we can write,

$$U(z,s) = \Psi(s)\mathrm{e}^{-\frac{z}{v}s}$$

Inverting,

$$\begin{aligned} E_x(z,t) &= \mathcal{L}^{-1}\left\{\Psi(s)\mathrm{e}^{-\frac{z}{v}s}\right\} \\ &= \begin{cases} \psi(t-\frac{z}{v}) & \text{for} \quad z < vt \\ 0 & \text{for} \quad z > vt \end{cases} \end{aligned}$$

□

Example 5.30. Repeat the Example 5.29 with the additional constraint that there is a PEC at $z = \ell$.

Solution: With a PEC at $z = \ell$, we have an additional constraint $U(\ell, s) = 0$.

Applying boundary conditions to the general solution

$$U(z,s) = A(s)\mathrm{e}^{-\frac{z}{v}s} + B(s)\mathrm{e}^{\frac{z}{v}s}$$

we get

$$\begin{aligned} \Psi(s) &= A(s) + B(s) \\ 0 &= A(s)\mathrm{e}^{-\frac{\ell}{v}s} + B(s)\mathrm{e}^{\frac{\ell}{v}s} \end{aligned}$$

solving, which we get

$$\begin{aligned} A(s) &= \frac{\Psi(s)\mathrm{e}^{\ell s/v}}{\mathrm{e}^{\ell s/v} - \mathrm{e}^{-\ell s/v}} \\ B(s) &= -\frac{\Psi(s)\mathrm{e}^{-\ell s/v}}{\mathrm{e}^{\ell s/v} - \mathrm{e}^{-\ell s/v}} \end{aligned}$$

Now the solution for $U(z,s)$ is given as

$$\begin{aligned} U(z,s) &= \Psi(s)\frac{\mathrm{e}^{(\ell-z)s/v} - \mathrm{e}^{-(\ell-z)s/v}}{\mathrm{e}^{\ell s/v} - \mathrm{e}^{-\ell s/v}} \\ &= \Psi(s)\frac{\mathrm{e}^{-zs/v} - \mathrm{e}^{-(2\ell-z)s/v}}{1 - \mathrm{e}^{-2\ell s/v}} \end{aligned}$$

Taking inverse Laplace transform, we get spatio-temporal variation of the field. First let us denote $\mathcal{L}^{-1}\left\{\Psi(s)\left(\mathrm{e}^{-zs/v} - \mathrm{e}^{-(2\ell-z)s/v}\right)\right\} = \psi(t - z/v) - \psi(t + z/v - 2\ell/v)$ as $\widetilde{\psi}(z,t)$. We recognize that the required inverse is the causal periodic version of $\widetilde{\psi}(z,t)$. Hence,

$$E_x(z,t) = \sum_{k=0}^{\infty} \widetilde{\psi}(z, t - 2k\ell/v)$$

□

Example 5.31. Solve the initial-boundary value problem using Laplace transform

$$\begin{aligned}
\frac{\partial^2\phi}{\partial t^2} &= \frac{\partial^2\phi}{\partial x^2}, && 0 < x < a, \quad t > 0 \\
\phi(x,0) &= b\sin\frac{\pi x}{a}, && 0 < x < a \\
\phi_t(x,0) &= -b\sin\frac{\pi x}{a}, && 0 < x < a \\
\phi(0,t) &= 0, && t > 0 \\
\phi(a,t) &= 0, && t > 0
\end{aligned}$$

Using the Laplace transform, the partial differential equation and the boundary conditions become

$$\begin{aligned}
\frac{\mathrm{d}^2\Phi}{\mathrm{d}x^2} &= s^2\Phi - bs\sin\frac{\pi x}{a} + b\sin\frac{\pi x}{a} \\
\Phi(0,s) &= 0 \\
\Phi(a,s) &= 0
\end{aligned}$$

the solution of which can be written as

$$\Phi(x,s) = \left(\frac{a^2 b(s-1)}{a^2 s^2 + \pi^2}\right)\sin\frac{\pi x}{a}$$

Now taking inverse transform, we get

$$\phi(x,t) = b\sin\frac{\pi x}{a}\left[\cos\frac{\pi t}{a} - \frac{a}{\pi}\sin\frac{\pi t}{a}\right]$$

□

5.6 Conclusion

Integral transforms occupy a central place in the solution of boundary value problems in general. We have introduced some of the common integral transforms and showed how to solve a range of problems encountered in EM. It is interesting to note that in some cases we get a closed form solution by a systematic application of the transforms. A future challenge in this topic is to develop the integral transforms in discrete version and the associated fast algorithms to obtain solutions on a digital computer.

References

[1] B. Davies, *Integral Transforms and Their Applications,* New York: Springer, 2002.

[2] L. C. Andrews and B. K. Shivamoggi, *Integral Transforms for Engineers,* Bellingham: SPIE Optical Engineering Press, 1999.

[3] L. C. Andrews, *Elementary Partial Differential Equations with Boundary Value Problems,* Orlando, FL: Academic Press, 1986, pp. 171–172, 429–435.

[4] R. E. Collin, *Field Theory of Guided Waves,* New York: IEEE Press, 1991.

[5] N. N. Lebedev, I. P. Skalskaya, and Y. S. Uflyand, *Worked Problems in Applied Mathematics,* New York: Dover Publications, 1965.

[6] I. N. Sneddon, *The Use of Integral Transforms,* New York: McGraw-Hill, 1972.

[7] H. J. Eom, *Electromagnetic Wave Theory for Boundary Value Problems,* New York: Springer, 2004.

Problems

1. Solve the Laplace equation $\nabla^2 V = 0$ to find the potential $V(x, y)$ in the region between the infinite planar boundaries $y = 0$ and $y = d$ that are maintained at

$$\begin{aligned} V(x,0) &= p(x) \\ V(x,d) &= q(x) \end{aligned}$$

Use the Fourier transform method.

2. The infinite planar boundaries $y = 0$ and $y = d$ have an infinitesimal gap at $x = \pm h$. They are maintained at a potential given by

$$V(x,0) = V(x,d) \quad = \quad V_0\left[(x+h) - u(x-h)\right]$$

where $u(x)$ is the Heaviside unit step function. Show that solving the Laplace equation $\nabla^2 V = 0$ for the potential $V(x, y)$, one gets

$$V(x,y) \quad = \quad \frac{V_0}{d}\sin\left(\frac{\pi y}{d}\right)\int_{-h}^{h}\frac{\cosh\pi(x-t)dt}{\left[\cosh^2\frac{\pi}{d}(x-t) - \cos^2\left(\frac{\pi y}{d}\right)\right]}$$

3. Show that

$$\sum_{n=1}^{\infty} \epsilon_n \cos nx = 2\pi \sum_{n=-\infty}^{\infty} \delta(x - 2n\pi)$$

where

$$\epsilon_n = \begin{cases} 1 & n = 1 \\ 2 & n > 1 \end{cases}$$

and $\delta(\cdot)$ is the Dirac-delta function.

4. Solve the equation

$$\left(\nabla^2 + k^2\right) A_z = -\mu\delta(x)\delta(y)\delta(z^2 - h^2)$$

for A_z with the constraint that $A_z \to 0$ as $\sqrt{x^2 + y^2 + z^2} \to \infty$.

5. Solve the equation

$$\left(\nabla^2 + k^2\right) A_z = \mu\delta(x)\delta(y)\left[\delta(z + h) - \delta(z - h)\right]$$

for A_z with the constraint that $A_z \to 0$ as $\sqrt{x^2 + y^2 + z^2} \to \infty$.

6. Solve

$$u_{tt} - c^2 u_{xx} = \mathrm{e}^{-2t} \sin x$$

subject to

$$u(x, 0) = u_t(x, 0) = 0$$

over the domain $-\infty < x < \infty, \quad t > 0$.

7. Solve

$$u_{tt} - c^2 u_{xx} = 0$$

subject to

$$u(x, 0) = u_t(x, 0) = 0, \quad u_x(0, t) = t, \quad u_x(\ell, t) = 0$$

over the domain $0 < x < \ell, \quad t > 0$.

8. Use the Hankel transform to solve

$$u_{tt} - c^2 \left(u_{\rho\rho} + \frac{1}{\rho} u_\rho\right) = 0$$

subject to

$$u(\rho, 0) = \frac{1}{\sqrt{1 + \rho^2}}; \quad u_t(\rho, 0) = 0$$

over the domain $\rho > 0, \quad t > 0$.

9. Use the Hankel transform to solve

$$u_{tt} + a^2 u_t - c^2\left(u_{\rho\rho} + \frac{1}{\rho}u_\rho\right) = 0$$

subject to

$$u(\rho, 0) = p(\rho); \quad u_t(\rho, 0) = q(\rho)$$

over the domain $\rho > 0, \quad t > 0$.

10. Show the following Mellin transforms:

(a) $\mathcal{M}\left[e^{-a\rho^2}\right] = \frac{a^{-\frac{k}{2}}(k/2-1)!}{2}$

(b) $\mathcal{M}\left[\frac{1+\rho\cos\phi}{1+2\rho\cos\phi+\rho^2}\right] = \frac{\pi\cos k\phi}{\sin\pi k}$

11. Find the Mellin transform of the cosine integral $\mathrm{Ci}(\rho) = -\int_\rho^\infty \frac{\cos t}{t}dt$.

12. Consider the wedge-shaped domain: $0 \le \rho < \infty, \quad |\varphi| \le \alpha$. The radial planes have an infinitesimal slit at the radius $\rho = a$. The finite corner is held at 1 volt while the infinite portions are grounded as given by

$$V(\rho, \pm\alpha) = \begin{cases} 1, & 0 \le \rho < a \\ 0, & \rho > a \end{cases}$$

Show that the electric potential $V(\rho, \varphi)$ in the entire space between the planes is given by:

$$V(\rho, \varphi) = \begin{cases} 1 - \frac{1}{\pi}\tan^{-1}\left[\frac{2(a\rho)^\beta \cos\beta\varphi}{a^{2\beta} - \rho^{2\beta}}\right], & 0 \le \rho < a \\ \frac{1}{\pi}\tan^{-1}\left[\frac{2(a\rho)^\beta \cos\beta\varphi}{\rho^{2\beta} - a^{2\beta}}\right], & \rho > a \end{cases}$$

where $\beta = \frac{\pi}{2\alpha}$.

13. Consider the wave equation in the half-space $x \ge 0, \quad -\infty \le y \le \infty$

$$\nabla^2\psi = \frac{1}{c^2}\frac{\partial^2\psi}{\partial t^2}$$

with the initial and boundary conditions

$$\begin{aligned} \psi(x, y, 0) &= 0 \\ \psi_t(x, y, 0) &= 0 \\ \psi(0, y, t) &= g(y) \end{aligned}$$

Show that the solution can be written as

$$\psi(x, y, t) = \begin{cases} 0 & t < \frac{x}{c} \\ \frac{xct}{\pi}\int_{y-y'}^{y+y'} \frac{g(\alpha)}{\rho^2\sqrt{c^2t^2 - \rho^2}}d\alpha & t > \frac{x}{c} \end{cases}$$

where $y' = \sqrt{c^2t^2 - x^2}$ and $\rho = \sqrt{x^2 + (y - \alpha)^2}$. Hint: Use the Laplace and Fourier transforms in the variables x and y, respectively.

14. Consider a lossless transmission line of length ℓ with a characteristic impedance Z_0 lying along the x-axis terminated by a resistive load of R_L at $x = \ell$. Suppose a time varying voltage source of $v_{in}(t)$ is directly connected to the terminals at the point $x = 0$. Assume that phase velocity along the line is denoted by u. Let $V_{in}(s)$ be the Laplace transform of $v_{in}(t)$. Show that the Laplace transform of the voltage $v(x, t)$ at a general point x is given by

$$\begin{aligned} \mathcal{L}\{v(x,t)\} &= V(x,s) \\ &= V_{in}(s)\left[\frac{R_L \cosh\left((\ell - x)\frac{s}{u}\right) + Z_0 \sinh\left((\ell - x)\frac{s}{u}\right)}{R_L \cosh\left(\frac{s\ell}{u}\right) + Z_0 \sinh\left(\frac{s\ell}{u}\right)}\right] \end{aligned}$$

15. Suppose a capacitor C initially charged to a voltage V_0 is connected to the line in the preceding problem at $t = 0$ as shown in Figure 5.18. Determine $V(x, s)$.

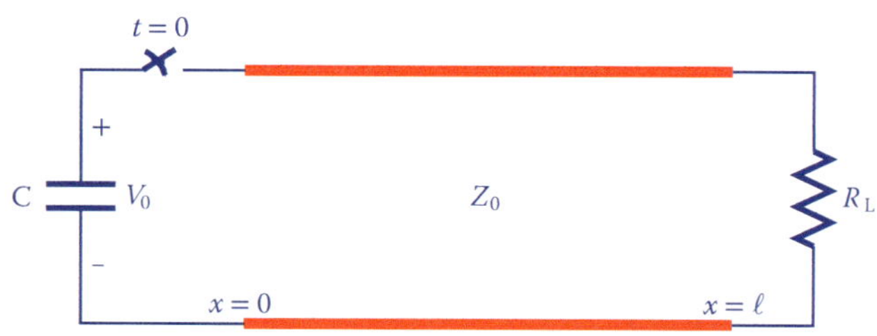

FIGURE 5.18
A capacitor C initially charged to a voltage of V_0 is connected to a load resistance R_L through a lossless line of length ℓ.

6

Perturbation Methods

When a man sits with a pretty girl for an hour it seems like a minute. But let him sit on a hot stove for a minute and its longer than any hour. That's relativity.

– Albert Einstein

CONTENTS

6.1 Introduction

The analytical methods dealt in the preceding chapters offer exact solutions to boundary value electromagnetic problems with such geometries as are easily tractable at least up to the stage we prepare the solution in some form. In some cases the solutions are in closed form, easy for numerical calculation and in other cases, the solutions are in either integral form or in a series form. Not all geometries are easily incorporated in deriving the solution. Sometimes a small deviation in a problem that has an a priori known exact solution could necessitate going through the rough road all over again or sometimes it may not be easy to formulate the problem and make it ready for solving. In such cases, we can start with the original problem called the "unperturbed problem" and proceed to solve the "perturbed problem" to some degree of accuracy. Perturbation methods are a class of methods that can yield approximate solutions that are in the neighborhood of unperturbed counterparts of the original problems [1–2].

In most cases, when a small (or large) parameter p occurs in the equations that characterize the problem, the solution is expanded in a series of powers of p, (or $\frac{1}{p}$) and this series is back-substituted and a system of equations is generated by coefficient matching technique. Then obtaining the sought-for parameter p is found to be easier through this system than through the original formulation of the problem. The series used in this process sometimes converges quickly. Then we call the perturbation method. But in other cases, we may need to add more and more terms thereby allowing the solution to converge in an asymptotic way. Then the technique is called asymptotic method. The guidelines to be followed for carrying out the expansion are not evident in every case. As such, it is expedient to scrutinize the exact solutions of some standard canonical problems to gain insight [3–6].

In the first part of this chapter, we work out some problems that provide us a foretaste of how a perturbation scheme offers a reasonably accurate solution to an otherwise harder problem. We shall start with a quadratic equation whose roots are known, but we pretend as if we do not know them and apply the technique to see what accuracy we get. Then we shall also solve the Helmholtz equation in a cylindrical region that has a known analytic solution, but we shall attempt an approximation. Following this we will obtain flux plot and equipotential plot in case of a dented perfect electric conductor (PEC) cylinder immersed in a uniform electric field. Then we try a nonlinear problem and compare the result with a more accurate numerical answer.

In the next part of the chapter, we provide some examples. Also we address the problem of finding resonant frequencies of electromagnetic cavities due to shape and material perturbations.

6.2 The Underlying Technique

Although the problems amenable for solution using perturbation techniques are diverse in nature, the underlying strategy can be succinctly explained in abstract terms as follows [3–4]: Suppose we are given an operator equation

$$\mathcal{T}f = g \tag{6.1}$$

which is not malleable to solve using the known methods. But we wish it was of the form

$$\mathcal{L}f = g \tag{6.2}$$

where $\mathcal{L}$ is a well-known linear operator whose solution is routine and systematic. By linear, we mean that for any arbitrary constants α and β

$$\mathcal{L}(\alpha f_1 + \beta f_2) = \alpha\mathcal{L}f_1 + \beta\mathcal{L}f_2 \tag{6.3}$$

Now, we rewrite the original problem in Equation 6.1 as

$$\begin{aligned}\mathcal{L}f - \mathcal{L}f + \mathcal{T}f &= g\\ \mathcal{L}f &= g + (\mathcal{L}f - \mathcal{T}f)\\ \mathcal{L}f &= g + \mathcal{A}f\end{aligned}$$

where $\mathcal{A}f = (\mathcal{L}f - \mathcal{T}f)$. Now, we consider a slightly generalized functional equation

$$\mathcal{L}f = g + \epsilon\mathcal{A}f \tag{6.4}$$

wherein we used a perturbation parameter ϵ, which when equals zero, the solution of Equation 6.4 is the known solution of $\mathcal{L}f = g$ denoted by f_0. But when $\epsilon = 1$, we get the sought for solution of Equation 6.1. However, as ϵ goes from 0 to zero in "bits and pieces," we obtain an insight into how the known solution undergoes the required modification. Let us now assume that this modification can be expressed as a series in the positive powers of ϵ as

$$f = f_0 + \epsilon f_1 + \epsilon^2 f_2 + \ \ldots \tag{6.5}$$

Operating on f by $\mathcal{L}$, we see that

$$\begin{aligned}\mathcal{L}f &= \mathcal{L}(f_0 + \epsilon f_1 + \epsilon^2 f_2 + \ \ldots)\\ &= \mathcal{L}f_0 + \epsilon\mathcal{L}f_1 + \epsilon^2\mathcal{L}f_2 + \ldots\end{aligned} \tag{6.6}$$

Also let us assume that the analytic nature of $\mathcal{A}$ permits us to write

$$\begin{aligned}\mathcal{A}f &= \mathcal{A}(f_0 + \epsilon f_1 + \epsilon^2 f_2 + \ \ldots)\\ &= \mathcal{A}(f_0) + \epsilon\mathcal{A}_1(f_0, f_1) + \epsilon^2\mathcal{A}_2(f_0, f_1, f_2) + \ldots\\ g + \epsilon\mathcal{A}f &= g + \epsilon\mathcal{A}(f_0) + \epsilon^2\mathcal{A}_1(f_0, f_1) + \epsilon^3\mathcal{A}_2(f_0, f_1, f_2) + \ldots\end{aligned} \tag{6.7}$$

Next, equating like powers of ϵ in both Equations 6.6 and 6.7, we generate the following system:

$$\begin{aligned}\mathcal{L}(f_0) &= g\\ \mathcal{L}(f_1) &= \mathcal{A}(f_0)\\ \mathcal{L}(f_2) &= \mathcal{A}_1(f_0, f_1)\\ &\vdots\\ \mathcal{L}(f_{n+1}) &= \mathcal{A}_k(f_0, \ldots, f_n)\\ &\vdots\end{aligned} \tag{6.8}$$

This is an infinite system of equations that can be solved recursively term after term that compose the needed solution. Of course, if $\epsilon = 1$, we get exact answer. But if ϵ is a fraction, the series might hopefully converge to a reasonably satisfactory answer. Let us work out few examples and illustrate this method.

Example 6.1. Find the roots of the quadratic equation: $f(x) := x^2 - 5x + 6 + \epsilon = 0$ using perturbation method. In particular, for $\epsilon = 0.1$, compare the approximate roots with the exact ones.

Solution: Let us assume that the dependence of each root can be described in a power series in ϵ. Assuming a_n to be the nth coefficient, substituting $x = \sum_{n=0}^{\infty} a_n \epsilon^n$ in the original quadratic, we get

$$f(x, \epsilon) = \sum_{n=0}^{\infty} c_n \epsilon^n \tag{6.9}$$

where

$$\begin{aligned} c_1 &= 1 - 5a_1 + 2a_0 a_1 \\ c_2 &= -5a_2 + 2a_0 a_2 + a_1^2 \\ c_3 &= -5a_3 + 2a_0 a_3 + 2a_1 a_2 \\ c_4 &= 2a_0 a_4 + 2a_1 a_3 + a_2^2 - 5a_4 \\ &\quad \cdots \end{aligned} \tag{6.10}$$

Equating each of the above to zero permits us to express a's as

$$\begin{aligned} a_1 &= \frac{1}{5 - 2a_0} \\ a_2 &= \frac{a_1^2}{5 - 2a_0} \\ a_3 &= \frac{2a_1 a_2}{5 - 2a_0} \\ a_4 &= \frac{2a_1 a_3 + a_2^2}{5 - 2a_0} \\ &\quad \cdots \end{aligned} \tag{6.11}$$

which can be determined recursively with the knowledge of a_0 to start with. Using the values of a's thus found, we can write the two roots α and β of the original quadratic equation as

$$\begin{aligned} \alpha &= 2 + \epsilon + \epsilon^2 + 2\epsilon^3 + 5\epsilon^4 + \mathcal{O}(\epsilon^5) \\ \beta &= 3 - \epsilon + \epsilon^2 - 2\epsilon^3 + 5\epsilon^4 + \mathcal{O}(\epsilon^5) \end{aligned} \tag{6.12}$$

Substituting $\epsilon = 0.1$, we get

$$\begin{aligned} \alpha \approx & \ 2.1125 \\ \beta \approx & \ 2.9085 \end{aligned} \tag{6.13}$$

whereas the exact roots are $5/2 - \sqrt{1/4 - \epsilon} = 2.1127$ and $5/2 - \sqrt{1/4 + \epsilon} = 2.8873$. □

Example 6.2. Consider a wave function in cylindrical coordinates $\psi(\rho)$ that varies only with ρ and satisfies the Helmholtz equation

$$\nabla^2 \psi + k^2 \psi = 0 \tag{6.14}$$

inside the cylindrical region $\rho < 1$ with the boundary condition $\psi(1) = 1$. Solve for $\psi(\rho)$ using perturbation technique and compare your answer with the analytical solution.

Solution: Let us expand the sought for function in the power series of k^2 as

$$\psi = \sum_{n=0}^{\infty} \psi_n k^{2n} \tag{6.15}$$

Let us insert this in the original differential equation 6.14

$$\begin{aligned} \nabla^2 \sum_{n=0}^{\infty} \psi_n k^{2n} + k^2 \sum_{n=0}^{\infty} \psi_n k^{2n} &= 0 \\ \nabla^2 \left\{ \psi_0 + \sum_{n=1}^{\infty} \psi_n k^{2n} \right\} + \sum_{n=0}^{\infty} \psi_n k^{2(n+1)} &= 0 \\ \nabla^2 \left\{ \psi_0 + \sum_{n=1}^{\infty} \psi_n k^{2n} \right\} + \sum_{n=1}^{\infty} \psi_{n-1} k^{2n} &= 0 \\ \nabla^2 \psi_0 + \sum_{n=1}^{\infty} \left\{ \nabla^2 \psi_n + \psi_{n-1} \right\} k^{2n} &= 0 \end{aligned} \tag{6.16}$$

Equating each coefficient of k^{2n} to zero, we get

$$\begin{aligned} \nabla^2 \psi_0 = & \quad 0 \\ \nabla^2 \psi_n = & \ -\psi_{n-1} \end{aligned} \tag{6.17}$$

Inserting the series in the boundary condition, we get

$$\begin{aligned} \psi(1) = \sum_{n=0}^{\infty} \psi_n(1) k^{2n} = & \quad 1 \\ \Rightarrow \qquad \psi_0(1) = & \quad 1 \\ \psi_n(1) = & \ 0 \qquad n \geq 1 \end{aligned} \tag{6.18}$$

Note that the perturbation method replaced the original Helmholtz equation by a system of equations, the first of which, is a Laplace's equation and the remaining are Poisson's equations to be solved in a sequential way. Let us proceed solving them.

$$\begin{aligned} \nabla^2\psi_0 = \frac{1}{\rho}\frac{\mathrm{d}}{\mathrm{d}\rho}\left(\rho\frac{\mathrm{d}\psi_0}{\mathrm{d}\rho}\right) &= 0 \\ \Rightarrow \qquad \rho\frac{\mathrm{d}\psi}{\mathrm{d}\rho} &= C_1\ln\rho + C_2 \end{aligned} \tag{6.19}$$

A solution that remains bounded for $\rho = 0$ and that satisfies $\psi_0(1) = 1$ is simply $\psi_0 = 1$. Now let us solve for a bounded ψ_1 such that $\psi_1(1) = 0$.

$$\begin{aligned} \nabla^2\psi_1 = \frac{1}{\rho}\frac{\mathrm{d}}{\mathrm{d}\rho}\left(\rho\frac{\mathrm{d}\psi_1}{\mathrm{d}\rho}\right) &= -1 \\ \Rightarrow \qquad \psi_1 &= \frac{1-\rho^2}{4} \end{aligned} \tag{6.20}$$

Now combining the two functions ψ_0 and ψ_1, we can write the solution as

$$\psi(\rho) = 1 + \left(\frac{1-\rho^2}{4}\right)k^2 + \mathrm{O}(k^4) \tag{6.21}$$

The given Helmholtz equation when ψ is a function of ρ alone has the standard solution bounded inside the cylindrical region in the form of Bessel functions given by

$$\psi(\rho) = \frac{J_0(k\rho)}{J_0(k)} \tag{6.22}$$

Figure 6.1 gives a plot of ψ for $k = 0.4$ obtained from Equations 6.21 and 6.22. □

Example 6.3. An infinite cylinder $\rho = a$ made of a PEC is immersed in a y-directed uniform electric field of E_0V/m. Draw the flux lines and the equipotential lines. Modify your answer if the cylinder is distorted and is now described by the perturbed equation $\rho(\varphi, \epsilon) = a(1 - \epsilon\sin^2\varphi)$. Take a value of $\epsilon = 0.5$. Assume also that the cylinder is grounded.

Solution: First, let us consider C, the regular PEC cylinder $\rho = a$ placed in the electric field $\mathbf{E} = E_0\hat{y}$. The potential is zero at the origin and inside the cylinder because it is grounded. At larger distances away from the cylinder the field is essentially $\mathbf{E} = E_0\hat{y}$ and hence the potential is $V(\rho,\varphi) = -E_0 y = -E_0\rho\sin\varphi$. From our knowledge of series solution, we might construct a potential function in the form

$$V(\rho,\varphi) = \sum_{k=1}^{\infty}\left(A_k\rho^k + \frac{B_k}{\rho^k}\right)(C_k\cos k\varphi + D_k\sin k\varphi) \tag{6.23}$$

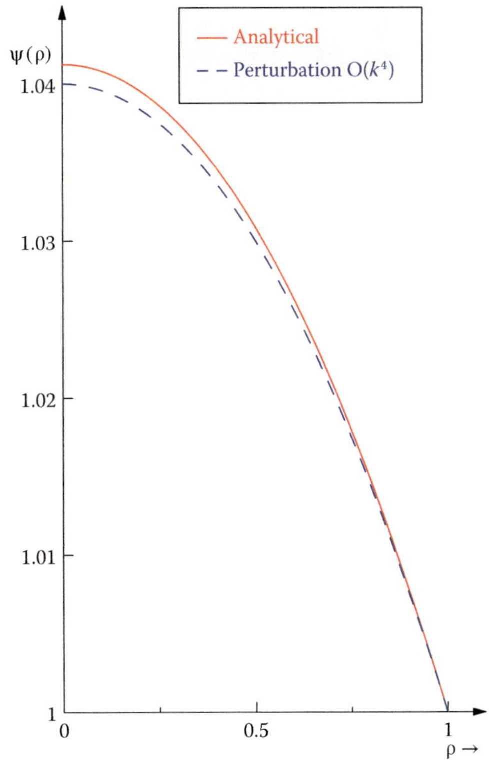

FIGURE 6.1
Solution of $(\nabla^2 + k^2)\psi = 0$ for $k = 0.4$ inside the cylindrical region $\rho < 1$ with the boundary condition $\psi(1) = 1$ using analytical and perturbation methods.

In order that the potential be zero on the cylinder and far away from the cylinder it approaches the expression $-E_0\rho\sin\varphi$, we choose only $k = 1$ term and write the potential in case of the unperturbed cylinder as

$$V(\rho, \varphi) \quad = \quad -E_0\left(\rho - \frac{a^2}{\rho}\right)\sin\varphi \tag{6.24}$$

Now coming to the case of perturbed cylinder, say C′, given by $\rho = a(1 - \epsilon\sin^2\varphi)$, substituting this in Equation 6.24 and writing as power series in ϵ

$$\begin{aligned} V(a, \varphi)|_{C'} &= \left(2E_0 a\sin^3\varphi\right)\epsilon + \mathcal{O}(\epsilon^2) \\ &= \epsilon\left(\frac{3aE_0}{2}\sin\varphi - \frac{aE_0}{2}\sin 3\varphi\right) + \mathcal{O}(\epsilon^2) \end{aligned}$$

This offers an insight into how we can rework the solution for the perturbed case. Let us denote the potential in case of the perturbed cylinder by $V'(\rho, \varphi)$. If we desire an accuracy of $\mathcal{O}(\epsilon^2)$, we need to consider three terms in Equation 6.23 and seek an expansion of the form

$$V'(\rho, \varphi) \quad = \quad -E_0 \left(\rho - \frac{a^2}{\rho}\right) \sin k\varphi + \epsilon \left[\sum_{k=1}^{3} \left(\frac{B'_k}{\rho^k}\right) \sin k\varphi\right]$$

We need to find B'_k such that $V(a, \varphi)|_{C'} = -V'(a, \varphi)$. After some simplification we get, $B'_1 = -\frac{3a^2 E_0}{2}, B'_2 = 0, B'_3 = \frac{a^4 E_0}{2}$. We are now in a position to express the potential in case of C$'$ as

$$V'(\rho, \varphi) \quad = \quad -E_0 \left(\rho - \frac{a^2}{\rho}\right) \sin k\varphi - \epsilon \frac{E_0}{2} \left[\frac{3a^2}{\rho} \sin \varphi - \frac{a^4}{\rho^3} \sin 3\varphi\right] + \mathcal{O}(\epsilon^2)$$

The equipotential lines and the flux lines are shown in Figure 6.2.

□

Laplace's equation was solved over the cross section of a rectangular trough in Section 2.2. Suppose the top plate is slightly tilted, the potential can be found by perturbation method as follows. A formal example is given below.

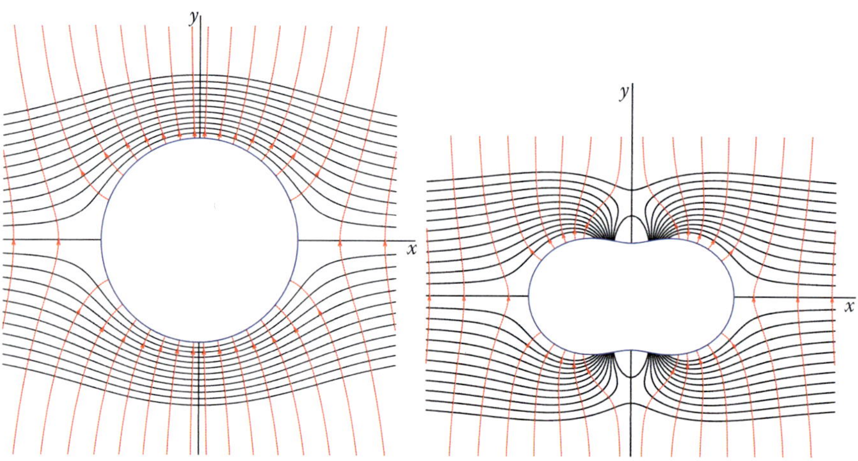

FIGURE 6.2
An infinite perfect electric conductor cylinder in a uniform electric field without and with perturbation in shape. These two cases are described by $\rho = a$ and $\rho = a(1 - \epsilon \sin^2 \phi)$, respectively. The equipotentials are in black and the flux lines are in red with arrows. The plot is shown for $a = 1$ and $\epsilon = 0.5$.

Example 6.4. Consider a trapezoidal region [4] in the xy-plane inside which electrostatic potential satisfies the Laplace equation and the boundary conditions as shown in Figure 6.3.

Solution: Let us write the solution for $V(x, y)$ as a sum of power series in ϵ as

$$V(x,y) = \sum_{n=0}^{\infty} \epsilon^n V_n(x,y)$$

$$\nabla^2 V(x,y) = 0 \Rightarrow \begin{cases} \nabla^2 V_0(x,y) = 0 \\ \nabla^2 V_1(x,y) = 0 \\ \cdots \quad \cdots \end{cases}$$

Along the top plate $y = b + \epsilon x$ the potential can be written using Taylor series as

$$\begin{aligned} V(x, b+\epsilon x) &= V(x,b) + \epsilon x V_y(x,b) + O(\epsilon^2) \\ &= f(x) \end{aligned}$$

Substituting the power series,

$$f(x) = \sum_{n=0}^{\infty} V_n(x,b)\epsilon^n + \epsilon x \sum_{n=0}^{\infty} V_{ny}(x,b)\epsilon^n + O(\epsilon^2)$$

$$\Rightarrow \begin{cases} V_0(x,b) = f(x) \\ V_1(x,b) = -xV_0(x,b) \end{cases}$$

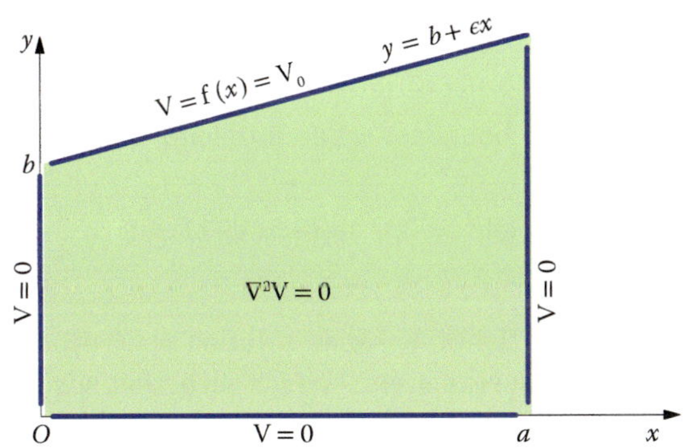

FIGURE 6.3
Cross section of a trapezoidal conducting trough.

Thus, the original problem is broken into two subproblems for $O(\epsilon^2)$ accuracy as

$$\left\{\begin{array}{c}\nabla^2 V(x,y)=0\\ V(x,0)=0\\ V(0,y)=0\\ V(a,y)=0\\ V(x,a+\epsilon x)=f(x)\end{array}\right\} \Rightarrow \left\{\begin{array}{c}\nabla^2 V_0(x,y)=0\\ V_0(x,0)=0\\ V_0(0,y)=0\\ V_0(a,y)=0\\ V_0(x,b)=f(x)\end{array}\right\} + \left\{\begin{array}{c}\nabla^2 V_1(x,y)=0\\ V_1(x,0)=0\\ V_1(0,y)=0\\ V_1(a,y)=0\\ V_1(x,b)=-xV_{0y}(x,b)\end{array}\right\}$$

We can express the solution for $V_0(x,y)$ from example in Section 2.2.1 as

$$V_0(x,y) = \frac{4V_T}{\pi}\sum_{n=1,3,5,\ldots}\frac{\sin\frac{n\pi x}{a}\sinh\frac{n\pi y}{a}}{n\sinh\frac{n\pi b}{a}}$$

Differentiating the above partially with respect to y and evaluating the same at $y=b$,

$$\begin{aligned}V_{0y}(x,b) &= \frac{4V_T}{\pi}\sum_{n=1,3,5,\ldots}\frac{n\pi}{a}\frac{\sin\frac{n\pi x}{a}\cosh\frac{n\pi b}{a}}{n\sinh\frac{n\pi b}{a}}\\ &= \frac{4V_T}{a}\sum_{n=1,3,5,\ldots}\sin\frac{n\pi x}{a}\coth\frac{n\pi b}{a}\\ V_1(x,y) &= \sum_{n=1}^{\infty}C_n\sinh\frac{n\pi y}{a}\sin\frac{n\pi x}{a}\end{aligned}$$

where

$$C_n = \frac{2}{a\sinh\frac{n\pi b}{a}}\int_0^a[-xV_{oy}(x,a)]\sin\frac{n\pi x}{a}dx$$

Thus, combining expressions for $V_0(x,y)$ and $\epsilon V_1(x,y)$, we get formal solution for the present problem. □

Example 6.5. Assuming that ϵ is very small, using perturbation method solve the one-dimensional boundary value problem:

$$\frac{\partial^2\psi}{\partial x^2}+\epsilon\psi^2=0,\quad \psi(0)=\psi(1)=1$$

Solution: Let us attempt a series solution $\psi(x,\epsilon) = \sum_{k=0}^{\infty}\psi_k(x)\epsilon^k$ wherein the individual functions are chosen such that $\psi_k(0)=\psi_k(1)=0$ for $k\geq 1$ and $\psi_0(0)=\psi_0(1)=1$. Back substitution yields

$$\frac{\partial^2\psi_0}{\partial x^2}+\sum_{k=1}^{\infty}\left(\frac{\partial^2\psi_k}{\partial x^2}+\sum_{m=0}^{k-1}\psi_m\psi_{k-m-1}\right)\epsilon^k = 0$$

In order that the power series in ϵ be zero, we enforce each term to zero. First we get

$$\begin{aligned} \frac{\partial^2 \psi_0}{\partial x^2} &= 0, \qquad \psi_0(0) = \psi_0(1) = 1, \\ \Rightarrow \quad \psi_0 &= 1 \end{aligned}$$

Next we get

$$\begin{aligned} \frac{\partial^2 \psi_1}{\partial x^2} &= -\psi_0^2 \qquad \psi_1(0) = \psi_1(1) = 0 \\ \Rightarrow \quad \psi_1 &= \frac{1}{2}x(1-x) \end{aligned}$$

In the same way,

$$\begin{aligned} \frac{\partial^2 \psi_1}{\partial x^2} &= -2\psi_0\psi_1 \qquad \psi_1(0) = \psi_1(1) = 0 \\ \Rightarrow \quad \psi_2 &= \frac{x}{12}(x-1)(x^2-x-1) \end{aligned}$$

and so on. Combining we get

$$\psi(x,\epsilon) = 1 + \frac{1}{2}\epsilon x(1-x) + \frac{\epsilon^2}{12}x(x-1)(x^2-x-1) + \mathcal{O}(\epsilon^3) \quad (6.24a)$$

For a choice of $\epsilon = 0.1$ Equation 6.24a is solved numerically and both the numerical and perturbation solutions are plotted in Figure 6.4.

□

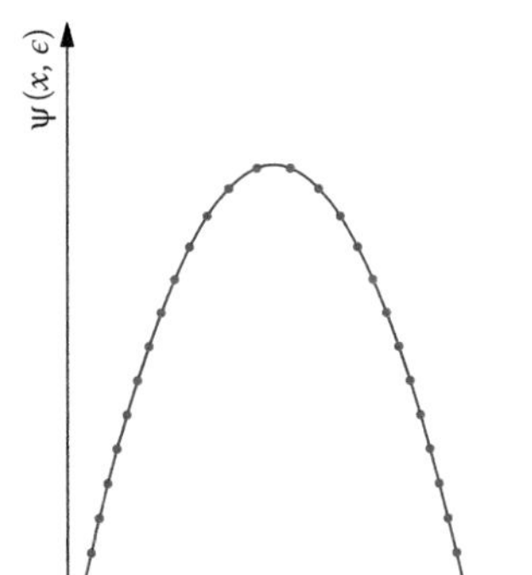

FIGURE 6.4
Numerical (solid) and perturbation (dotted) solutions of the boundary value problem: $\psi'' + 0.1\psi^2 = 0, \psi(0) = \psi(1) = 1$. It is hard to distinguish the two!

6.3 Electromagnetic Cavities

In this section, we discuss how perturbational methods can be employed in obtaining resonant frequency of cavities. Part of the material presented here is adapted from S. R. Nelatury and C. F. Nelatury [5]. Finding exact analytical formulas for characteristic parameters of electromagnetic resonators is possible only for certain regular shapes. If there are any aberrations either in shape or in the material medium filling the cavity, perturbation, and/or variational techniques are usually employed. When the degree of perturbation is small, the formulas obtained via these techniques give reasonably good approximation for the desired quantity; but if the perturbation is large, the accuracy is compromised.

Most of the perturbation or variational formulas are based on mathematical identities from vector calculus. Perturbation methods are supposed to work well for smaller changes. We take the unperturbed problem and use the pertinent fields as the trial fields needed to plug in the perturbation formulas. The choice of trial fields and to what extent of the boundary they satisfy the required boundary conditions determine the accuracy.

6.3.1 Formulas Based on Vector Identities

Consider an arbitrary resonant cavity formed by a closed conducting surface. Let S and τ denote its boundary and the volume, respectively. Let the cavity suffer a perturbation in shape and let S' and τ' denote the boundary of the perturbed cavity and its volume. The changes in the surface and volume are denoted by ΔS and $\Delta\tau$. We shall assume that the medium filling the cavity is lossless, linear, homogeneous, and isotropic with permittivity ϵ and permeability μ. See Figure 6.5. Suppose, $\mathbf{E_0}$, $\mathbf{H_0}$, ω_0 denote the fields and the resonant frequency of the unperturbed resonator, respectively, and let $\mathbf{E}, \mathbf{H}, \omega$ be the corresponding quantities of the perturbed cavity.

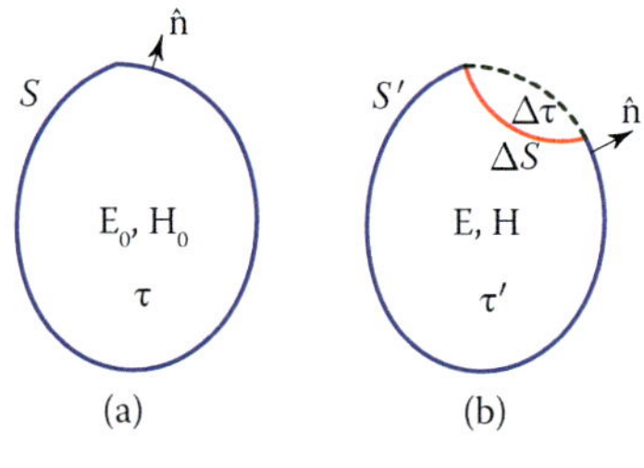

FIGURE 6.5
Perturbation of cavity walls. (*a*) Original cavity. (*b*) Perturbed cavity.

The fields in the unperturbed and perturbed cases satisfy

$$-\nabla \times \mathbf{E_0} = j\omega_0\mu\mathbf{H_0} \tag{6.25}$$
$$-\nabla \times \mathbf{E} = j\omega\mu\mathbf{H} \tag{6.26}$$
$$\nabla \times \mathbf{H_0} = j\omega_0\epsilon\mathbf{E_0} \tag{6.27}$$
$$\nabla \times \mathbf{H} = j\omega\epsilon\mathbf{E} \tag{6.28}$$

Performing dot multiplication on both sides of Equation 6.28 with $\mathbf{E_0^*}$ and the conjugate of Equation 6.25 with $\mathbf{H}$ we get

$$\mathbf{E_0^*} \cdot \nabla \times \mathbf{H} = j\omega\epsilon\mathbf{E_0^*} \cdot \mathbf{E} \tag{6.29}$$
$$-\mathbf{H} \cdot \nabla \times \mathbf{E_0^*} = -j\omega_0\mu\mathbf{H} \cdot \mathbf{H_0^*} \tag{6.30}$$

Now adding the above and applying the vector identity

$$\nabla \cdot (\mathbf{A} \times \mathbf{B}) \equiv \mathbf{B} \cdot \nabla \times \mathbf{A} - \mathbf{A} \cdot \nabla \times \mathbf{B}$$

we have

$$\nabla \cdot (\mathbf{H} \times \mathbf{E_0^*}) = j\omega\epsilon\mathbf{E_0^*} \cdot \mathbf{E} - j\omega_0\mu\mathbf{H} \cdot \mathbf{H_0^*} \tag{6.31}$$

Analogously, if we do the same with Equations 6.26 and 6.27 we have

$$\nabla \cdot (\mathbf{H_0^*} \times \mathbf{E}) = j\omega\mu\mathbf{H} \cdot \mathbf{H_0^*} - j\omega_0\epsilon\mathbf{E_0^*} \cdot \mathbf{E} \tag{6.32}$$

We can add Equations 6.31 and 6.32 and integrate throughout the volume of the perturbed cavity. In so doing, if we use the divergence theorem and the fact that $\mathbf{n} \times \mathbf{E} = \mathbf{0}$ on S' we get

$$\oiint_{S'} (\mathbf{H} \times \mathbf{E_0^*}) \cdot d\mathbf{S} = j(\omega - \omega_0) \iiint_{\tau'} (\epsilon\mathbf{E} \cdot \mathbf{E_0^*} + \mu\mathbf{H} \cdot \mathbf{H_0^*})\, d\tau. \tag{6.33}$$

Since $\mathbf{n} \times \mathbf{E_0} = \mathbf{0}$ on S, the left-hand side can be written as

$$\begin{aligned}\oiint_{S'} (\mathbf{H} \times \mathbf{E_0^*}) \cdot d\mathbf{S} &= \oiint_{S'-S} (\mathbf{H} \times \mathbf{E_0^*}) \cdot d\mathbf{S} \\ &= -\oiint_{\Delta S} (\mathbf{H} \times \mathbf{E_0^*}) \cdot d\mathbf{S}\end{aligned} \tag{6.34}$$

We can now rewrite Equation 6.33 as

$$\omega - \omega_0 = \frac{j \oiint_{\Delta S} (\mathbf{H} \times \mathbf{E_0^*}) \cdot d\mathbf{S}}{\iiint_{\tau'} (\epsilon\mathbf{E} \cdot \mathbf{E_0^*} + \mu\mathbf{H} \cdot \mathbf{H_0^*})\, d\tau} \tag{6.35}$$

From the above, an exact expression and an approximation to it for the ratio $\frac{\omega}{\omega_0}$ can be written as

$$\frac{\omega}{\omega_0} = \left[1 + \frac{j \oiint_{\Delta S} (\mathbf{H} \times \mathbf{E}_0^*) \cdot d\mathbf{S}}{\omega_0 \iiint_{\tau'} (\epsilon \mathbf{E} \cdot \mathbf{E_0}^* + \mu \mathbf{H} \cdot \mathbf{H_0}^*) \, d\tau}\right] \tag{6.36}$$

$$\approx \left[1 + \frac{j \oiint_{\Delta S} (\mathbf{H_0} \times \mathbf{E}_0^*) \cdot d\mathbf{S}}{\omega_0 \iiint_{\tau'} (\epsilon |E_0|^2 + \mu |H_0|^2) \, d\tau}\right] \tag{6.37}$$

The approximation is made here by replacing $\mathbf{E}$, $\mathbf{H}$ by the unperturbed fields $\mathbf{E_0}$, $\mathbf{H_0}$.

6.3.2 Formulas Based on Variational Technique

Variational formulas are slightly different from perturbation formulas, in the sense that they yield an approximate expression to a sought for quantity, whose first variation with respect to perturbations vanishes. Such formulas are said to be *stationary*. We shall consider two variational expressions for the square of the resonant frequency which can be obtained from the wave equations:

$$\nabla \times \mu^{-1} \nabla \times \mathbf{E} = \omega^2 \epsilon \mathbf{E} \tag{6.38}$$

$$\nabla \times \epsilon^{-1} \nabla \times \mathbf{H} = \omega^2 \mu \mathbf{H} \tag{6.39}$$

Performing scalar multiplication by $\mathbf{E}$ on both sides of Equation 6.38 and integrating over the volume of the cavity

$$\omega^2 = \frac{\iiint_{\tau'} \mathbf{E} \cdot \nabla \times \mu^{-1} \nabla \times \mathbf{E} \, d\tau}{\iiint_{\tau'} \epsilon E^2 \, d\tau} \tag{6.40}$$

Similarly, scalar multiplication by $\mathbf{H}$ on both sides of Equation 6.39 and integrating over the volume of the cavity

$$\omega^2 = \frac{\iiint_{\tau'} \mathbf{H} \cdot \nabla \times \epsilon^{-1} \nabla \times \mathbf{H} \, d\tau}{\iiint_{\tau'} \mu H^2 \, d\tau} \tag{6.41}$$

To assess the stationarity of Equation 6.40 or Equation 6.41, a trial field in terms of an arbitrary parameter α is substituted in place of the true field. For instance, let us work with Equation 6.40 and substitute $\mathbf{E}_{\text{trial}} = \mathbf{E} + \alpha \mathbf{e}$ in place of $\mathbf{E}$ and find derivative of ω^2 with respect to α at $\alpha = 0$. Denoting the numerator of the right-hand side of Equation 6.40 by $N(\alpha)$ and its denominator by $D(\alpha)$, we get

$$\left.\frac{\partial \omega^2}{\partial \alpha}\right|_{\alpha=0} = \frac{D(0)N'(0) - N(0)D'(0)}{D^2(0)} \tag{6.42}$$

$$= -\frac{\oiint_{S'} \left[\left(\mu^{-1} \nabla \times \mathbf{E} \right) \times \mathbf{e} \right] \cdot d\mathbf{S}}{\iiint_{\tau'} \epsilon E^2 d\tau} \tag{6.43}$$

which vanishes if $\mathbf{E}_{\text{trial}}$ is so chosen that the tangential component thereof on the cavity wall is zero. However, suppose we consider a correction term

$$c = \frac{\oiint_{S'} \left[\left(\mu^{-1} \nabla \times \mathbf{E} \right) \times \mathbf{E} \right] \cdot d\mathbf{S}}{\iiint_{\tau'} \epsilon E^2 d\tau} \tag{6.44}$$

and find the derivative as before at $\alpha = 0$, after some simplification we get

$$\left. \frac{\partial c}{\partial \alpha} \right|_{\alpha=0} = \frac{\oiint_{S'} \left[\left(\mu^{-1} \nabla \times \mathbf{E} \right) \times \mathbf{e} \right] \cdot d\mathbf{S}}{\iiint_{\tau'} \epsilon E^2 d\tau} \tag{6.45}$$

Hence, if Equation 6.40 is modified as

$$\omega^2 = \frac{\iiint_{\tau'} \mathbf{E} \cdot \nabla \times \mu^{-1} \nabla \times \mathbf{E}\, d\tau + \oiint_{S'} \left[\left(\mu^{-1} \nabla \times \mathbf{E} \right) \times \mathbf{E} \right] \cdot d\mathbf{S}}{\iiint_{\tau'} \epsilon E^2 d\tau} \tag{6.46}$$

it stays stationary irrespective of whether the trial field $\mathbf{E}$ satisfies boundary condition on the cavity walls. The first term in the numerator of Equation 6.46 can be expanded and we have finally

$$\omega^2 = \frac{\iiint \mu^{-1} \left(\nabla \times \mathbf{E} \right)^2 d\tau + 2 \oiint \left[\left(\mu^{-1} \nabla \times \mathbf{E} \right) \times \mathbf{E} \right] \cdot d\mathbf{S}}{\iiint \epsilon E^2 d\tau} \tag{6.47}$$

and hence,

$$\frac{\omega}{\omega_0} = \sqrt{\left[\frac{\iiint_{\tau'} \left| \nabla \times \mathbf{E} \right|^2 d\tau + 2 \oiint_{S'} \left(\nabla \times \mathbf{E} \right) \times \mathbf{E} \cdot d\mathbf{S}}{\iiint_{\tau'} k_0^2 E^2 d\tau} \right]} \tag{6.48}$$

where $k_0^2 = \omega_0^2 \mu \epsilon$. A similar approach helps us to obtain the perturbation formula in terms of $\mathbf{H}$ vector as

$$\frac{\omega}{\omega_0} = \sqrt{\left[\frac{\iiint_{\tau'} \left| \nabla \times \mathbf{H} \right|^2 d\tau}{\iiint_{\tau'} k_0^2 H^2 d\tau} \right]} \tag{6.49}$$

In the derivation of the above, we note that a trial field $\mathbf{H}$ is not required to satisfy any boundary conditions.

6.3.3 Slater's Perturbational Formula

Slater has derived a perturbational formula from the modal expansion of fields in a closed region. General solutions of wave equation valid in a spatial domain of *finite volume* form a linear space $\mathcal{U}$ spanned by a complete set of orthonormal vectors. More generally, we might view all conceivable electric and magnetic field solutions in such a finite region as forming a vector space

$\mathcal{U}$. This can be built as a direct sum (denoted by the symbol $\oplus$) of two subspaces of vectors, one spanned by a *countable* set of solenoidal orthonormal vectors $\{\mathbf{E}_n, \mathbf{H}_n\}$ with zero divergence and the other by a *countable* set of irrotational orthonormal vectors $\{\mathbf{F}_n, \mathbf{G}_n\}$, whose curl is the zero vector. We can write $\mathcal{U} = \{\mathbf{E}_n, \mathbf{H}_n\} \oplus \{\mathbf{F}_n, \mathbf{G}_n\}$. The vectors $\mathbf{E}_n, \mathbf{H}_n, \mathbf{F}_n, \mathbf{G}_n$ are called mode vectors and any expansion of an arbitrary vector field in terms of these is called modal expansion. We can expand the electric field in terms of $\{\mathbf{E}_n, \mathbf{F}_n\}$ and the magnetic field in terms of $\{\mathbf{H}_n, \mathbf{G}_n\}$. As long as the domain is *finite* as in the case of a cavity for example, the series expansion is *discrete*. If the domain is *infinite*, it becomes continuous and is represented by integration instead of discrete summation. An example is the whole three dimensional space, which usually come across in radiation problems. Accordingly, the fields $\mathbf{E}$ and $\mathbf{H}$ can be written as

$$\mathbf{E} = \sum_n \Big(\langle \mathbf{E}, \mathbf{E}_n \rangle \mathbf{E}_n + \langle \mathbf{E}, \mathbf{F}_n \rangle \mathbf{F}_n \Big) \tag{6.50}$$

$$\mathbf{H} = \sum_n \Big(\langle \mathbf{H}, \mathbf{H}_n \rangle \mathbf{H}_n + \langle \mathbf{H}, \mathbf{G}_n \rangle \mathbf{G}_n \Big) \tag{6.51}$$

where the inner product $\langle \mathbf{A}, \mathbf{B} \rangle$ is defined over the volume of the cavity τ as

$$\langle \mathbf{A}, \mathbf{B} \rangle = \iiint_\tau \mathbf{A} \cdot \mathbf{B} \, \mathrm{d}\tau \tag{6.52}$$

This induces a norm $\|\mathbf{A}\|$ for a vector $\mathbf{A}$ given by

$$\|\mathbf{A}\| = \sqrt{\langle \mathbf{A}, \mathbf{A} \rangle} \tag{6.53}$$

Orthonormality requires $\forall\, k, l$, and $\forall\, m \neq n$,

$$\left.\begin{array}{llll} \langle \mathbf{E}_k, \mathbf{F}_l \rangle & = \langle \mathbf{E}_k, \mathbf{H}_l \rangle = \langle \mathbf{E}_k, \mathbf{G}_l \rangle & = & 0 \\ \langle \mathbf{F}_k, \mathbf{H}_l \rangle & = \langle \mathbf{F}_k, \mathbf{G}_l \rangle = \langle \mathbf{H}_k, \mathbf{G}_l \rangle & = & 0 \\ \langle \mathbf{E}_m, \mathbf{E}_n \rangle & = \langle \mathbf{F}_m, \mathbf{F}_n \rangle & = & 0 \\ \langle \mathbf{H}_m, \mathbf{H}_n \rangle & = \langle \mathbf{G}_m, \mathbf{G}_n \rangle & = & 0 \\ \|\mathbf{E}_n\| = \|\mathbf{H}_n\| & = \|\mathbf{F}_n\| = \|\mathbf{G}_n\| & = & 1 \end{array}\right\} \tag{6.54}$$

It may be noted that in physical reality although $\mathbf{H}$ field is solenoidal, for the sake of mathematical completeness, we include the irrotational component, which vanishes in the exact solution. Instead of leaving the mode field vectors as arbitrary, we may make further assumptions that

$$\left.\begin{array}{lll} \nabla \times \mathbf{E}_n & = k_n \mathbf{H}_n & \\ \nabla \times \mathbf{H}_n & = k_n \mathbf{E}_n & \\ \hat{\mathbf{n}} \times \mathbf{E}_n & = \mathbf{0} & \text{on } S_s \\ \hat{\mathbf{n}} \times \mathbf{H}_n & = \mathbf{0} & \text{on } S_o \\ \hat{\mathbf{n}} \cdot \mathbf{E}_n & = 0 & \text{on } S_o \\ \hat{\mathbf{n}} \cdot \mathbf{H}_n & = 0 & \text{on } S_s \\ \hat{\mathbf{n}} \times \mathbf{F}_n & = \mathbf{0} & \text{on } S_s \cup S_o \\ \hat{\mathbf{n}} \cdot \mathbf{G}_n & = 0 & \text{on } S_s \cup S_o \end{array}\right\} \tag{6.55}$$

where k_n is a constant (identified as the wave number), $S_s \cup S_o$ forms the bounding surface of the volume τ, S_s is metallic and short circuits the electric field whereas, S_o is the remaining part that stays open.

In the same way, we can also expand the curl of electric and magnetic field vectors as

$$\nabla \times \mathbf{E} = \sum_n \Big(\langle \nabla \times \mathbf{E}, \mathbf{H}_n \rangle \mathbf{H}_n + \langle \nabla \times \mathbf{E}, \mathbf{G}_n \rangle \mathbf{G}_n \Big) \tag{6.56}$$

$$\nabla \times \mathbf{H} = \sum_n \Big(\langle \nabla \times \mathbf{H}, \mathbf{E}_n \rangle \mathbf{E}_n + \langle \nabla \times \mathbf{H}, \mathbf{F}_n \rangle \mathbf{F}_n \Big) \tag{6.57}$$

Let us consider an arbitrary cavity subjected to shape perturbation. Start with the identity

$$\begin{aligned} \nabla \cdot (\mathbf{E} \times \nabla \times \mathbf{E}_n) &= \nabla \times \mathbf{E} \cdot \nabla \times \mathbf{E}_n - \mathbf{E} \cdot \nabla \times \nabla \times \mathbf{E}_n \\ &= k_n \mathbf{H}_n \cdot \nabla \times \mathbf{E} - k_n^2 \, \mathbf{E} \cdot \mathbf{E}_n \end{aligned} \tag{6.58}$$

Integrating Equation 6.58 over the volume τ and making use of divergence theorem, we get

$$\begin{aligned} \iint_{S_s} \mathbf{E} \times \mathbf{H}_n \cdot d\mathbf{S} + \iint_{S_o} \mathbf{E} \times \mathbf{H}_n \cdot d\mathbf{S} &= \iiint_\tau (\nabla \times \mathbf{E} \cdot \mathbf{H}_n) \, \mathrm{d}\tau \\ &\quad - k_n \iiint_\tau \mathbf{E} \cdot \mathbf{E}_n \mathrm{d}\tau \end{aligned} \tag{6.59}$$

But because the tangential component of $\mathbf{H}_n$ is zero on S_o, integration over S_o vanishes and we exchange $(\mathbf{E} \times \mathbf{H}_n \cdot d\mathbf{S})$ with $(\mathbf{H}_n \cdot \hat{\mathbf{n}} \times \mathbf{E} \, \mathrm{d}S)$ and have

$$\begin{aligned} \iiint_\tau (\nabla \times \mathbf{E} \cdot \mathbf{H}_n) \, \mathrm{d}\tau &= \iint_{S_s} \mathbf{H}_n \cdot (\hat{\mathbf{n}} \times \mathbf{E}) \, \mathrm{d}S \\ &\quad + k_n \iiint_\tau \mathbf{E} \cdot \mathbf{E}_n \, \mathrm{d}\tau \end{aligned} \tag{6.60}$$

Another identity is apt for our purpose now:

$$\begin{aligned} \nabla \cdot (\mathbf{E} \times \mathbf{G}_n) &= \mathbf{G}_n \cdot \nabla \times \mathbf{E} - \mathbf{E} \cdot \nabla \times \mathbf{G}_n \\ &= \nabla \times \mathbf{E} \cdot \mathbf{G}_n \end{aligned} \tag{6.61}$$

Integrating Equation 6.61 over the volume τ

$$\begin{aligned} \iiint_\tau (\nabla \times \mathbf{E} \cdot \mathbf{G}_n) \, \mathrm{d}\tau &= \iint_{S_s} \hat{\mathbf{n}} \times \mathbf{E} \cdot \mathbf{G}_n \, \mathrm{d}S \\ &\quad + \iint_{S_o} \hat{\mathbf{n}} \times \mathbf{E} \cdot \mathbf{G}_n \, \mathrm{d}S \end{aligned} \tag{6.62}$$

As a similar exercise, we can perform integrations involved in the expansion of $\nabla \times \mathbf{H}$

$$\iiint\limits_{\tau} (\nabla \times \mathbf{H} \cdot \mathbf{E}_n)\, \mathrm{d}\tau = \iint\limits_{S_o} \mathbf{H} \cdot (\mathbf{E}_n \times \hat{\mathbf{n}})\, \mathrm{d}S + k_n \iiint\limits_{\tau} \mathbf{H} \cdot \mathbf{H}_n\, \mathrm{d}\tau \tag{6.63}$$

Using the above expressions, we can write

$$\nabla \times \mathbf{H} = \sum_n \mathbf{E}_n \left[k_n \langle \mathbf{H}, \mathbf{H}_n \rangle + \iint\limits_{S_o} \hat{\mathbf{n}} \times \mathbf{H} \cdot \mathbf{E}_n \mathrm{d}S \right] \tag{6.64}$$

$$\nabla \times \mathbf{E} = \sum_n \mathbf{H}_n \left[k_n \langle \mathbf{E}, \mathbf{E}_n \rangle \iint\limits_{Ss} \hat{\mathbf{n}} \times \mathbf{E} \cdot \mathbf{H}_n \mathrm{d}S \right] + \sum_n \mathbf{G}_n \iint\limits_{S_s \cup S_o} \hat{\mathbf{n}} \times \mathbf{E} \cdot \mathbf{G}_n\, \mathrm{d}S \tag{6.65}$$

Substituting these expressions into Maxwell's equation $\nabla \times \mathbf{E} = -\mu \frac{\partial \mathbf{H}}{\partial t}$, we get

$$\sum_n \mathbf{H}_n \left[k_n \langle \mathbf{E}, \mathbf{E}_n \rangle + \iint\limits_{S_s} (\hat{\mathbf{n}} \times \mathbf{E}) \cdot \mathbf{H}_n\, \mathrm{d}S \right] + \sum_n \mathbf{G}_n \iint\limits_{S_s \cup S_o} (\hat{\mathbf{n}} \times \mathbf{E}) \cdot \mathbf{G}_n\, \mathrm{d}S + \mu \frac{\partial}{\partial t} \left(\sum_n \mathbf{H}_n \langle \mathbf{H}, \mathbf{H}_n \rangle + \sum_n \mathbf{G}_n \langle \mathbf{H}, \mathbf{G}_n \rangle \right) = 0 \tag{6.66}$$

Performing dot product with $\mathbf{H}_n$, integrating over the volume while applying orthogonality properties, we get

$$k_n \langle \mathbf{E}, \mathbf{E}_n \rangle + \mu \frac{\partial}{\partial t} \langle \mathbf{H}, \mathbf{H}_n \rangle = - \iint\limits_{S_s} (\hat{\mathbf{n}} \times \mathbf{E}) \cdot \mathbf{H}_n\, \mathrm{d}S \tag{6.67}$$

Now following the same treatment over the Maxwell equation $\nabla \times \mathbf{H} = \epsilon \frac{\partial \mathbf{E}}{\partial t}$ assuming no conduction current inside the cavity, we get

$$k_n \langle \mathbf{H}, \mathbf{H}_n \rangle - \epsilon \frac{\partial}{\partial t} \langle \mathbf{E}, \mathbf{E}_n \rangle = - \iint\limits_{S_o} (\hat{\mathbf{n}} \times \mathbf{H}) \cdot \mathbf{E}_n\, \mathrm{d}S \tag{6.68}$$

Combining Equations 6.67 and 6.68, we can get equations for $\langle \mathbf{E}, \mathbf{E}_n \rangle$ and $\langle \mathbf{H}, \mathbf{H}_n \rangle$

$$\mu\epsilon \frac{\partial^2}{\partial t^2} \langle \mathbf{E}, \mathbf{E}_n \rangle + k_n^2 \langle \mathbf{E}, \mathbf{E}_n \rangle = \mu \frac{\partial}{\partial t} \iint\limits_{S_o} (\hat{\mathbf{n}} \times \mathbf{H}) \cdot \mathbf{E}_n\, \mathrm{d}S - k_n \iint\limits_{S_s} (\hat{\mathbf{n}} \times \mathbf{E}) \cdot \mathbf{H}_n\, \mathrm{d}S \tag{6.69}$$

and

$$\mu\epsilon\frac{\partial^2}{\partial t^2}\langle \mathbf{H}, \mathbf{H}_n\rangle + k_n^2\langle \mathbf{H}, \mathbf{H}_n\rangle$$
$$= -\epsilon\frac{\partial}{\partial t}\iint\limits_{S_o} (\hat{\mathbf{n}} \times \mathbf{E}) \cdot \mathbf{H}_n \, \mathrm{d}S - k_n \iint\limits_{S_o} (\hat{\mathbf{n}} \times \mathbf{H}) \cdot \mathbf{E}_n \, \mathrm{d}S \tag{6.70}$$

If we assume that the boundary surface S_s is a PEC and S_o is an open aperture, the tangential component $\hat{\mathbf{n}} \times \mathbf{E}$ vanishes on S_s and likewise $\hat{\mathbf{n}} \times \mathbf{H}$ vanishes on S_o. This results in

$$\mu\epsilon\frac{\partial^2}{\partial t^2}\langle \mathbf{E}, \mathbf{E}_n\rangle + k_n^2\langle \mathbf{E}, \mathbf{E}_n\rangle = 0 \tag{6.71}$$

and

$$\mu\epsilon\frac{\partial^2}{\partial t^2}\langle \mathbf{H}, \mathbf{H}_n\rangle + k_n^2\langle \mathbf{H}, \mathbf{H}_n\rangle = 0 \tag{6.72}$$

implying that $\langle \mathbf{E}, \mathbf{E}_n\rangle$ and $\langle \mathbf{H}, \mathbf{H}_n\rangle$ in general have sinusoidal variation with an angular frequency ω satisfying $\omega^2\mu\epsilon = k_n^2$. That prompts us to replace the time derivative by $j\omega$ in phasor notation.

Slater considered a cavity made of a perfect conductor and subjected to perturbation of the wall shown in Figure 6.6. He focused attention on the small volume $\Delta\tau$ by which a part of cavity wall is pushed in and it becomes $\overline{S}$. So the volume $\Delta\tau$ is bounded by $S \cup \overline{S}$.

Consider the identity involving the fields of the unperturbed cavity,

$$\nabla \cdot (\mathbf{E}_0 \times \mathbf{H}_0) = \mathbf{H}_0 \cdot \nabla \times \mathbf{E}_0 - \mathbf{E}_0 \cdot \nabla \times \mathbf{H}_0 \tag{6.73}$$
$$= k_0\left(H_0^2 - E_0^2\right) \tag{6.74}$$

wherein the first two subequations of Equation 6.55 are employed. Integrating Equation 6.74 over the perturbed volume $\Delta\tau$ and using divergence theorem

$$\iint\limits_{S\cup\overline{S}} \hat{\mathbf{n}} \cdot (\mathbf{E}_0 \times \mathbf{H}_0)\, \mathrm{d}S = k_0 \iiint\limits_{\Delta\tau} \left(H_0^2 - E_0^2\right) \mathrm{d}\tau \tag{6.75}$$

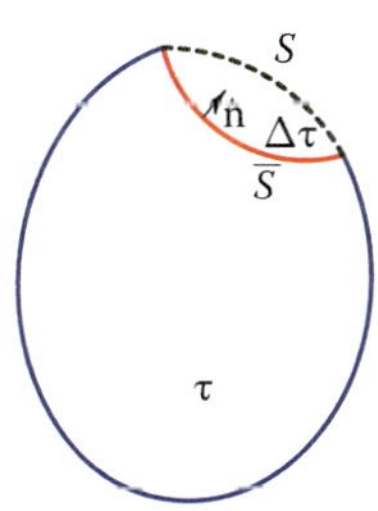

FIGURE 6.6
An arbitrary cavity with perturbed wall.

which, after striking off integration on S on the-left hand side, can be written as

$$\iint_{\overline{S}} (\hat{\mathbf{n}} \times \mathbf{H}_0) \cdot \mathbf{E}_0 \mathrm{d}S = k_0 \iiint_{\Delta\tau} \left(H_0^2 - E_0^2\right) \mathrm{d}\tau \tag{6.76}$$

Multiplying both sides by $k_0\langle \mathbf{H}, \mathbf{H}_0\rangle$

$$\iint_{\overline{S}} k_0 \left(\hat{\mathbf{n}} \times \langle \mathbf{H}, \mathbf{H}_0\rangle \mathbf{H}_0\right) \cdot \mathbf{E}_0 \mathrm{d}S = k_0^2 \langle \mathbf{H}, \mathbf{H}_0\rangle \iiint_{\Delta\tau} \left(H_0^2 - E_0^2\right) \mathrm{d}\tau \tag{6.77}$$

Now assuming that in the orthogonal expansion of the perturbed field $\mathbf{H}$, only the first term is predominant, that is, $\mathbf{H} \approx \langle \mathbf{H}, \mathbf{H}_0\rangle \mathbf{H}_0$, we can write

$$\iint_{\overline{S}} (\hat{\mathbf{n}} \times \mathbf{H}) \cdot \mathbf{E}_0 \mathrm{d}S = k_0 \langle \mathbf{H}, \mathbf{H}_0\rangle \iiint_{\Delta\tau} \left(H_0^2 - E_0^2\right) \mathrm{d}\tau \tag{6.78}$$

The left-hand side is component of the surface electric current along $\mathbf{E}_0$ on $\overline{S}$. Using Equation 6.70 we can write in phasor notation

$$\langle \mathbf{H}, \mathbf{H}_0\rangle \left(\omega^2 \mu\epsilon - k_0^2\right) = k_0^2 \langle \mathbf{H}, \mathbf{H}_0\rangle \iiint_{\Delta\tau} \left(H_0^2 - E_0^2\right) \mathrm{d}\tau \tag{6.79}$$

Hence,

$$\omega^2 \mu\epsilon = k_0^2 \left(1 + \iiint_{\Delta\tau} \left(H_0^2 - E_0^2\right) \mathrm{d}\tau\right) \tag{6.80}$$

from which Slater obtained a perturbation formula for the ratio $\frac{\omega}{\omega_0}$ as

$$\frac{\omega}{\omega_0} = \left[1 + \iiint_{\Delta\tau} \left(H_0^2 - E_0^2\right) \mathrm{d}\tau\right]^{\frac{1}{2}} \tag{6.81}$$

Slater's perturbational formula is based on expressing the component of electric surface current on the perturbed walls. The actual current distribution depends on the geometry and location of the perturbation and is related to the surface magnetic field, which has a discontinuity. If the sum of complete orthogonal components of a discontinuous function converges to 1, and if, on one side of the discontinuity, the function has a value 0, then it implies that on the other side it has a value of 2. With this in mind Dombrowski added a factor 2 in Slater's formula and that seems to make a significant difference. So according to this change the corrected version of Equation 6.81 now becomes

$$\frac{\omega}{\omega_0} = \left[1 + 2 \iiint_{\Delta\tau} \left(H_0^2 - E_0^2\right) \mathrm{d}\tau\right]^{\frac{1}{2}} \tag{6.82}$$

6.3.4 A Heuristic Perturbational Formula

Furthermore, the resonant frequency of a cavity can be expressed posing the Helmholtz equation satisfied by the wave function ψ

$$\nabla^2\psi + k^2\psi = 0 \tag{6.83}$$

as an eigen value problem. We might rewrite the wave equation as an operator equation

$$\mathscr{L}\psi := -\nabla^2\psi = \Lambda\psi \tag{6.84}$$

where $\Lambda = k^2$ is the eigen value of $\mathscr{L}$. Multiplying both sides by ψ and integrating over the given volume τ, we can obtain

$$\Lambda = \frac{\iiint_\tau \left(-\psi\nabla^2\psi\right) \mathrm{d}\tau}{\iiint_\tau \psi^2 \mathrm{d}\tau} \tag{6.85}$$

Using the vector identity $\nabla \cdot (\psi\nabla\psi) \equiv \psi\nabla \cdot \nabla\psi + |\nabla\psi|^2$ and divergence theorem, we might write Equation 6.85 as

$$\Lambda = \frac{\iiint_\tau |\nabla\psi|^2 \mathrm{d}\tau - \oiint_S \psi\nabla\psi \cdot d\mathbf{S}}{\iiint_\tau \psi^2 \mathrm{d}\tau} \tag{6.86}$$

If either ψ or its normal derivative vanish on the bounding surface S, and if the denominator has almost a constant value we observe that

$$\Lambda \propto \iiint\limits_\tau |\nabla\psi|^2 \mathrm{d}\tau \tag{6.87}$$

Based on the proportionality that an eigen value bears in Equation 6.87 and using the wave function valid for unperturbed case, a heuristic perturbation formula can be written as

$$\frac{\omega}{\omega_0} = \sqrt{\frac{\iiint_\tau |\nabla\psi|^2 \mathrm{d}\tau}{\iiint_{\tau'} |\nabla\psi|^2 \mathrm{d}\tau}} \tag{6.88}$$

6.3.5 A Square Cavity with Perturbed Corners

In order to assess the various perturbation formulas, it is expedient if we apply all of them on the same cavity; in the present case, we take a square cavity of side a, whose cross-section is shown in the inset of Figure 6.7. The dominant transverse magnetic (TM_{110}) fields may be obtained from the wave function ψ_{110}^{TM} given as

$$\psi_{110}^{\mathrm{TM}} = \sin\frac{\pi x}{a} \sin\frac{\pi y}{a} \tag{6.89}$$

The corresponding fields are

$$\mathbf{E} = -j\omega_0\mu \sin\frac{\pi x}{a}\sin\frac{\pi y}{a}\hat{z} \tag{6.90}$$

$$\mathbf{H} = \frac{\pi}{a}\left(\sin\frac{\pi x}{a}\cos\frac{\pi y}{a}\hat{x} - \cos\frac{\pi x}{a}\sin\frac{\pi y}{a}\hat{y}\right) \tag{6.91}$$

The resonant frequency for (TM_{110}) mode is given by

$$\omega_0 = \frac{\pi}{a}\sqrt{\frac{2}{\mu\epsilon}} \tag{6.92}$$

Next consider the perturbed cavity with all the corners symmetrically flattened. Let δ be the length of the flattened portion from each of the corners as shown in Figure 6.7. The center of the cavity cross-section in the $z = 0$ plane is chosen at $(a/2, a/2)$. The ratio (δ/a) represents the shape perturbation parameter and is denoted by p.

We shall substitute the field expressions from Equations 6.90 to 6.91 in Equations 6.37, 6.48, and 6.49. In Equation 6.88, we shall use Equation 6.89. After simplification, we get

$$A := \frac{\omega}{\omega_0} = \frac{\pi^2 - 2\pi^2 p^2 + 4\sin^2\pi p - \pi p \sin 2\pi p}{\pi^2 - 2\pi^2 p^2 + 2\sin^2\pi p} \tag{6.93}$$

$$B := \frac{\omega}{\omega_0} = \sqrt{\frac{\pi^2 - 2\pi^2 p^2 + 8\sin^2\pi p - 3\pi p \sin 2\pi p}{\pi^2 - 2\pi^2 p^2 + 4\sin^2\pi p - \pi p \sin 2\pi p}} \tag{6.94}$$

$$C := \frac{\omega}{\omega_0} = \sqrt{\frac{\pi^2 - 2\pi^2 p^2 + 4\sin^2\pi p - \pi p \sin 2\pi p}{\pi^2 - 2\pi^2 p^2 + \pi p \sin 2\pi p}} \tag{6.95}$$

$$D := \frac{\omega}{\omega_0} = \sqrt{\frac{\pi^2 + 4\sin^2\pi p - 2\pi p \sin 2\pi p}{\pi^2}} \tag{6.96}$$

$$E := \frac{\omega}{\omega_0} = \sqrt{\frac{\pi^2 + 8\sin^2\pi p - 4\pi p \sin 2\pi p}{\pi^2}} \tag{6.97}$$

$$F := \frac{\omega}{\omega_0} = \sqrt{\frac{\pi}{\pi - 2\pi p^2 + p \sin 2\pi p}} \tag{6.98}$$

where

$$p = \frac{\delta}{a} \tag{6.99}$$

Formulas A through F are plotted in Figure 6.7. A numerical solution for this problem can be readily obtained using the eigen mode solver that comes with the commercially available software *HFSS*. This plot is labeled as G in Figure 6.7. The resonant frequency of a cylindrical cavity of diameter a is given by $\frac{2x_{01}}{a\sqrt{\mu\epsilon}}$ where x_{01} stands for the first zero of the zeroth order Bessel function of the first kind, $J_0(x)$. As δ is increased, at some point the cross section of the cavity circumscribes the circle of diameter a. For this to happen, we select

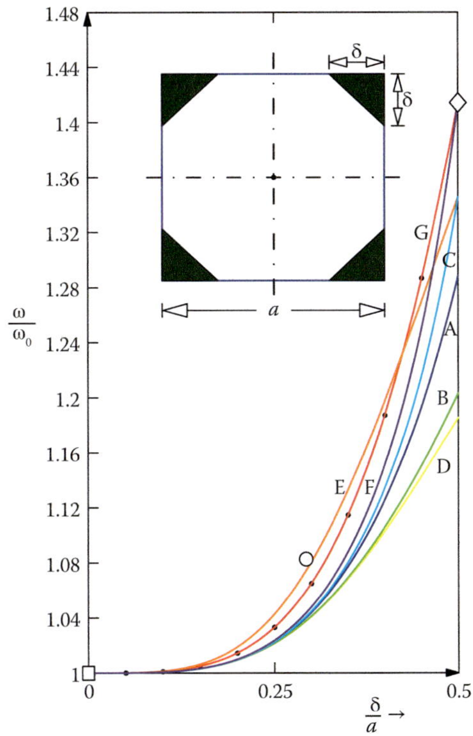

FIGURE 6.7
Comparison of plots based on various perturbation formulas in case of a square cavity. Inset shows the cavity cross-section. The three landmark cases whose exact frequencies are known are shown by markers resembling their cross-section.

$\delta = \frac{2-\sqrt{2}}{2}a$ or $p \approx 0.293$. We can label the three landmark cases by three symbols—square, circle, and diamond and write

$$\square := \left.\frac{\omega}{\omega_0}\right|_{p=0} = 1 \tag{6.100}$$

$$\bigcirc := \left.\frac{\omega}{\omega_0}\right|_{p=0.293} = \frac{x_{01}\sqrt{2}}{\pi} \tag{6.101}$$

$$\Diamond := \left.\frac{\omega}{\omega_0}\right|_{p=0.5} = \sqrt{2} \tag{6.102}$$

These markers are also shown in Figure 6.7. Note that the curve H passes through $\square$ and $\Diamond$ and is seen to be very close to $\bigcirc$ for the case of octagonal cross section, which is closer to being circular. All the curves, of course, pass through $\square$ as is expected for $p = 0$. All the seven plots are ranked based on how close they are to G. Treating H obtained through *HFSS* as reference, the error norm in each is calculated as: $\|e_D\| = 0.2879$, $\|e_B\| = 0.2691$, $\|e_A\| = 0.1741$, $\|e_C\| = 0.1188$, $\|e_E\| = 0.0760$, and $\|e_F\| = 0.0610$. In an increasing order of accuracy, we could rank the perturbation formulas A through H as

$$\mathrm{D} < \mathrm{B} < \mathrm{A} < \mathrm{C} < \mathrm{E} < \mathrm{F} < \mathrm{G} \tag{6.103}$$

Among A, B, and C, the latter is more accurate because the derivation of Equation 6.49 does not assume any boundary condition on **H** vector. Whereas, Equation 6.48 is made stationary by appending the second term in the numerator as a matter of mathematical necessity. Otherwise, the tangential component of the trial field is supposed to be zero on the boundary. The formula Equation 6.88, labeled F, is seen to be more accurate and its plot seems closest to G. It is consistent with the fact that square cavities of smaller side have greater resonant frequency. Further, the additional accuracy in F could be attributed to possibly balancing of approximately equal errors in the numerators and denominators.

6.4 Material Perturbations

In the preceding section, we showed how the resonant frequency of a cavity changes with shape perturbations. Let us now continue our discussion and see the effect of material perturbations on the resonant frequency of a cavity. Any change in the material filling the cavity brings about a change in the resonance. Also, a material body of certain dimensions and medium constants embedded in a cavity would cause a certain change in the resonant frequency thereof. Conversely, when a given material is embedded in a certain cavity, by measuring the net change in its resonant frequency, one could characterize the medium properties of the body inserted in the cavity. We can develop a perturbation formula for this correspondence and it can be done in a variety of ways just as we showed earlier. Toward this end, let us develop the needed mathematical steps. We shall include the method that Waldron proposed in 1960. Let us consider an empty cavity of volume τ enclosed by a closed surface S with the fields $\mathbf{E_0}$ and $\mathbf{H_0}$ resonating at the frequency ω_0. Suppose a material body of volume $\Delta\tau$ is introduced into the cavity resulting in the perturbed fields denoted by $\mathbf{E}$ and $\mathbf{H}$. Let the new resonance be ω. From Maxwell's equations,

$$\nabla \times \mathbf{E_0} = -j\omega_0 \mathbf{B}_0 \tag{6.104}$$

$$\nabla \times \mathbf{H} = j\omega \mathbf{D} \tag{6.105}$$

Performing dot multiplication on both sides of Equation 6.104 with **H**, and Equation 6.105 with $\mathbf{E_0}$ and subtracting the results we get

$$\mathbf{H} \cdot \nabla \times \mathbf{E_0} - \mathbf{E_0} \cdot \nabla \times \mathbf{H} = -j\omega_0 \mathbf{H} \cdot \mathbf{B}_0 - j\omega \mathbf{E_0} \cdot \mathbf{D} \tag{6.106}$$

Applying the vector identity

$$\nabla \cdot (\mathbf{A} \times \mathbf{B}) \equiv \mathbf{B} \cdot \nabla \times \mathbf{A} - \mathbf{A} \cdot \nabla \times \mathbf{B}$$

we have

$$\nabla \cdot (\mathbf{E_0} \times \mathbf{H}) \quad = \quad -j\omega_0 \mathbf{H} \cdot \mathbf{B}_0 - j\omega \mathbf{E_0} \cdot \mathbf{D} \tag{6.107}$$

Integrating Equation 6.107 over the volume of the cavity and using divergence theorem, we find that

$$\oiint_S (\mathbf{E}_0 \times \mathbf{H}) \cdot d\mathbf{S} = - \iiint_\tau (j\omega_0 \mathbf{H} \cdot \mathbf{B}_0 - j\omega \mathbf{E_0} \cdot \mathbf{D})\, \mathrm{d}\tau \tag{6.108}$$

A repeated operation with subscripts swapped yields

$$\oiint_S (\mathbf{E} \times \mathbf{H_0}) \cdot d\mathbf{S} = - \iiint_\tau (j\omega \mathbf{H_0} \cdot \mathbf{B} - j\omega_0 \mathbf{E} \cdot \mathbf{D}_0)\, \mathrm{d}\tau \tag{6.109}$$

Assuming a perfect conducting boundary for the cavity we find that the left-hand members of Equations 6.108 and 6.109 each must be equal to zero and hence we require that the right-hand sides must be equal to each other. This gives

$$j\omega_0 \iiint_\tau (\mathbf{E} \cdot \mathbf{D}_0 - \mathbf{H} \cdot \mathbf{B}_0)\, \mathrm{d}\tau = j\omega \iiint_\tau (\mathbf{E_0} \cdot \mathbf{D} - \mathbf{H_0} \cdot \mathbf{B})\, \mathrm{d}\tau \tag{6.110}$$

Suppose the perturbation is small, we expect a small change in the frequency; so, let us denote $\omega = \omega_0 + \Delta\omega$. Now, we can use this in Equation 6.110 and obtain

$$\frac{\Delta\omega}{\omega_0} = \frac{\iiint_\tau [(\mathbf{E} \cdot \mathbf{D}_0 - \mathbf{E_0} \cdot \mathbf{D}) - (\mathbf{H} \cdot \mathbf{B}_0 - \mathbf{H_0} \cdot \mathbf{B})]\, \mathrm{d}\tau}{\iiint_\tau (\mathbf{E_0} \cdot \mathbf{D} - \mathbf{H_0} \cdot \mathbf{B})\, \mathrm{d}\tau} \tag{6.111}$$

Here, we make an assumption that the perturbed and unperturbed field quantities differ in the region occupied by the material inclusion, but they are almost identical outside the material. (In fact, they differ even in the immediate vicinity of the object, but we ignore this.) Then, we can assume that the numerator of Equation 6.111 is zero outside the material and rewrite it as

$$\frac{\Delta\omega}{\omega_0} = \frac{\iiint_{\Delta\tau} [(\mathbf{E} \cdot \mathbf{D}_0 - \mathbf{E_0} \cdot \mathbf{D}) - (\mathbf{H} \cdot \mathbf{B}_0 - \mathbf{H_0} \cdot \mathbf{B})]\, \mathrm{d}\tau}{\iiint_\tau (\mathbf{E_0} \cdot \mathbf{D} - \mathbf{H_0} \cdot \mathbf{B})\, \mathrm{d}\tau} \tag{6.112}$$

If we make an assumption that the material perturber is made of a homogenous isotropic material of permittivity ϵ_r and permeability μ_r, we can rewrite Equation 6.112 as

$$\frac{\Delta\omega}{\omega_0} = \frac{\iiint_{\Delta\tau} [\epsilon_0 (1 - \epsilon_r) \mathbf{E_0} \cdot \mathbf{E} - \mu_0 (1 - \mu_r) \mathbf{H_0} \cdot \mathbf{H}]\, \mathrm{d}\tau}{\iiint_\tau (\mathbf{E_0} \cdot \mathbf{D} - \mathbf{H_0} \cdot \mathbf{B})\, \mathrm{d}\tau} \tag{6.113}$$

Further assumptions that the material is nonmagnetic and has $\mu_r = 1$, and that the perturbed fields in Equation 6.113 are approximated by the unperturbed ones, we can trim it as

$$\frac{\Delta\omega}{\omega_0} = \frac{\epsilon_0 (1 - \epsilon_r) \mathbf{E_0} \cdot \mathbf{E}}{4W_0} \Delta\tau \tag{6.114}$$

where W_0 is the energy in the unperturbed cavity. This arises from recognizing the denominator of Equation 6.113 as $4W_0$. Next we shall provide some simulation examples to validate the accuracy of Equation 6.114.

Example 6.6. Consider an air-filled cylindrical cavity of radius a, height ℓ, and dominant resonant frequency of ω_0 [6]. If the cavity is perturbed by inserting a dielectric rod of radius b and of the same length coaxially, causing the frequency to change to ω, determine and plot the ratio ω/ω_0 versus b/a using the perturbation formula Equation 6.114. Repeat the exercise using the analytical formula Equation 6.112 and compare them both for three sets of values $\epsilon_r = 3, 6$, and 12.

Solution: The given situation is depicted in Figure 6.8.

The volume of the rod is $\pi b^2 \ell$. Since the tangential component of the electric field stays continuous, we can approximate the perturbed electric field with the unperturbed version and rewrite Equation 6.114 as

$$\frac{\Delta\omega}{\omega_0} = \frac{\epsilon_0 (1 - \epsilon_r) |E_{00}|^2}{4W_0} \pi b^2 \ell$$

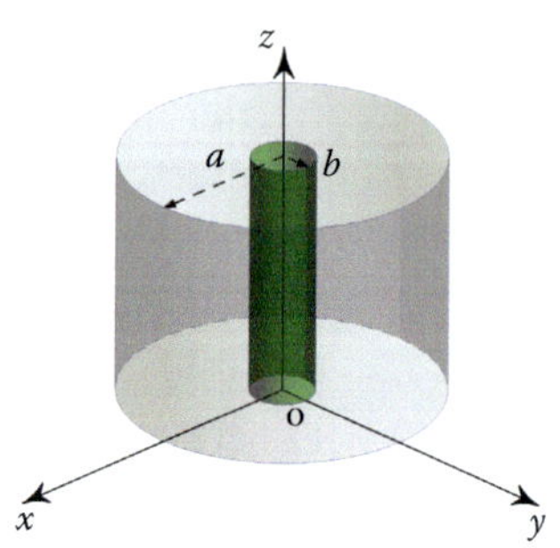

FIGURE 6.8
A cylindrical cavity coaxially perturbed by insertion of a dielectric rod.

We can take the z-directed electric field in the unperturbed cavity as $E_0 = |E_{00}|J_0(k_0\rho) = J_0(k_0\rho)$, assuming $|E_{00}| = 1$. This helps us to find the energy W_0 as

$$\begin{aligned} W_0 &= \int_{\rho=0}^{a}\int_{\varphi=0}^{2\pi}\int_{z=0}^{\ell} \frac{1}{2}\epsilon_0 E_0^2 \rho d\rho d\varphi dz \\ &= \frac{1}{2}\pi\epsilon_0 \ell a^2 \left(J_0^2(k_0 a) + J_1^2(k_0 a)\right) \end{aligned}$$

For the dominant mode taking $k_0 a \approx 2.4048$, we get

$$\frac{\Delta\omega}{\omega_0} \approx 1.8552\,(1-\epsilon_r)\,\frac{b^2}{a^2} \tag{6.115}$$

Next, in order to obtain an exact analytical formula, we need to express the fields in the perturbed cavity both inside and outside the dielectric rod. Let us construct the electric field as

$$E_z = \begin{cases} J_0(\sqrt{\epsilon_r}k_1\rho) & \text{for} \quad \rho < b \\ C_1 J_0(k_1\rho) + C_2 Y_0(k_1\rho) & \text{for} \quad b < \rho < a \end{cases}$$

where C_1 and C_2 are arbitrary constants, and k_1 is the wave number of the perturbed cavity. These are to be found from the boundary conditions:

$$\begin{aligned} E_z\big|_{\rho=a} &= 0 \\ E_z\big|_{\rho=b^-} &= E_z\big|_{\rho=b^+} \\ \frac{\partial E_z}{\partial \rho}\bigg|_{\rho=b^-} &= \frac{\partial E_z}{\partial \rho}\bigg|_{\rho=b^+} \end{aligned}$$

By enforcing the boundary conditions, we determine the arbitrary constants as

$$\begin{aligned} C_1 &= \frac{Y_0(k_0 a)J_0(\sqrt{\epsilon_r}k_1 b)}{J_0(k_1 b)Y_0(k_1 a) - J_0(k_1 a)Y_0(k_1 b)} \\ C_2 &= \frac{J_0(\sqrt{\epsilon_r}k_1 b)J_0(k_1 a)}{J_0(k_1 b)Y_0(k_1 a) - J_0(k_1 a)Y_0(k_1 b)} \end{aligned}$$

Further, we get the following transcendental equation:

$$-\sqrt{\epsilon_r}J_1(\sqrt{\epsilon_r}k_1 b) = \frac{J_0(\sqrt{\epsilon_r}k_1 b)\left[J_1(k_1 b)Y_0(k_1 a) - J_0(k_1 a)Y_1(k_1 b)\right]}{\left[J_0(k_1 b)Y_0(k_1 a) - J_0(k_1 a)Y_0(k_1 b)\right]}$$

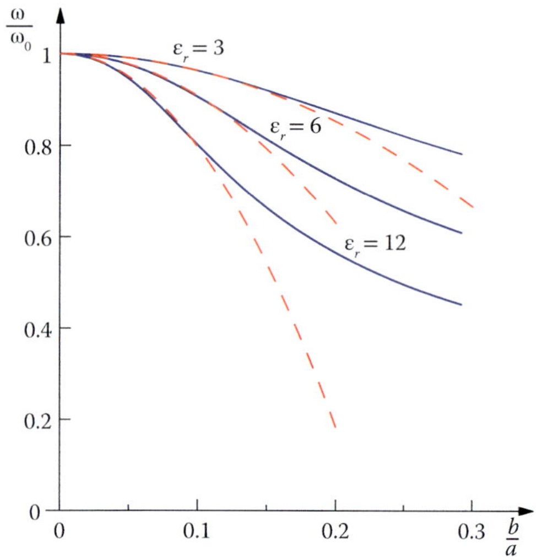

FIGURE 6.9
Comparison of the resonant frequency of a circular cavity perturbed by the introduction of a dielectric rod. Plots are shown for three different values of ϵ_r. Solid curve is based on exact formula, dashed curve is obtained from the perturbation formula given in Equation 6.115.

By solving Equation 6.116 numerically we can find k_1 and hence the perturbed frequency exactly. Likewise the perturbation formula Equation 6.115 yields an approximation. Figure 6.9 shows variation in ω/ω_0 as a function of b/a for three requested values of ϵ_r. □

6.5 Conclusion

The perturbational methods are a class of analytical methods that offer an approximate solution to a problem that is slightly different from another problem whose solution is known. The original problem whose solution is known beforehand is called *unperturbed* problem. Just because perturbational methods yield approximate solutions like the numerical methods, they cannot, however, supplant each other; but in real life, they should be used to supplement and confirm each other. The answers obtained by perturbation methods reflect the dependance of one or more crucial parameters in an intuitively satisfying manner. It is amazing that one who is conversant with these methods begins to view every problem as a perturbed version of another known problem. The scope of perturbation techniques extends far beyond linear systems and is now playing key role in the nonlinear regime also.

References

[1] R. F. Harrington, *Time-Harmonic Electromagnetic Fields,* New York: Wiley-Interscience, IEEE Press, 2001.

[2] R. E. Collin, *Field Theory of Guided Waves,* New York: IEEE Press, 1991.

[3] R. Bellman, *Perturbation Techniques in Mathematics, Engineering & Physics*, New York: Dover Publications, 1966.

[4] E. Zauderer, *Partial Differential Equations of Applied Mathematics,* New York: John Wiley & Sons, 1983.

[5] S. R. Nelatury and C. F. Nelatury, "A comparison of perturbation formulas for a square electromagnetic resonator," *IEEE Antennas Propagat. Mag.,* vol. 56, no. 1, pp. 130–142, Feb. 2014.

[6] R. G. Carter, "Accuracy of microwave cavity perturbation measurements," *IEEE Trans. Microwave Theory Tech.,* vol. 49, no. 5, pp. 918–923, May 2001.

Problems

1. Consider two concentric spheres of radii a and b with $b > a$. Let the space between them be filled with a dielectric of permittivity ϵ.

   (a) Show that the capacitance C_0 between them is

   $$C_0 = \frac{4\pi\epsilon ab}{b-a}$$

   (b) Now suppose the inner sphere is slightly displaced (perturbed) as shown in Figure 6.10. Show that an approximate expression for the potential in between is

   $$V = \frac{Q}{4\pi\epsilon}\left\{\frac{1}{r} - \frac{1}{b} + \frac{\delta r}{b^3 - a^3}\left[1 - \frac{a^3}{r^3}\right]\cos\theta\right\}$$

   where Q is the total charge on the inner sphere, r is the radial distance and θ is the elevation angle measured from its center. Assume that the outer sphere is grounded to begin your analysis.

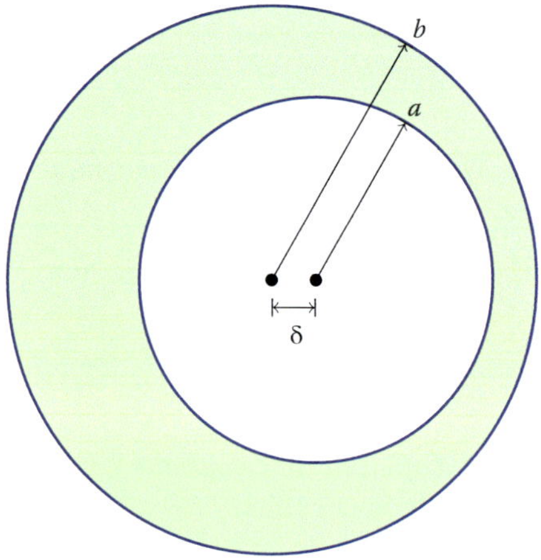

FIGURE 6.10
Two spheres perturbed from being concentric by a small parameter δ.

(c) Also show that the capacitance of the perturbed spheres is given by

$$C = C_0 \left\{ 1 + \frac{ab\delta^2}{(b-a)^2(b^2+ba+a^2)} \right\}$$

2. Two parallel square conducting plates of side a are separated in air by a small distance d. We know that the capacitance between them ignoring fringing is $C_0 = \frac{\epsilon a^2}{d}$. Suppose the top plate is slightly tilted as shown in Figure 6.11. Show that the capacitance can be approximated as

$$C = C_0 \left\{ 1 - \frac{a\vartheta}{2d} \right\}$$

3. A hollow waveguide of uniform cross-section is placed along the z-axis. The dielectric material filling the waveguide is isotropic but nonhomogeneous in that the dielectric permittivity is a function of z.

(a) Show that

$$\nabla^2 \mathbf{E} + \beta^2 \mathbf{E} = -\nabla \left(E_z \frac{\partial}{\partial z} \ln \epsilon \right)$$

(b) Suppose the functional expression for the permittivity is given by

$$\epsilon = \epsilon_{\text{ave}} + \Delta\epsilon(z)$$

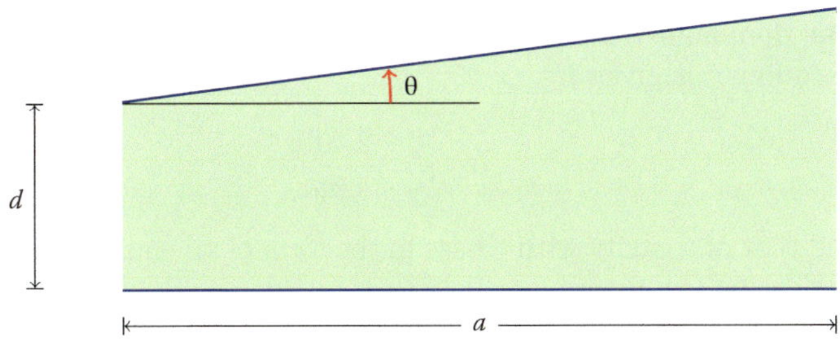

FIGURE 6.11
Two conducting plates originally parallel, but the top plate is slightly tilted by a small angle ϑ.

where $|\Delta\epsilon(z)| \ll \epsilon_{\text{ave}}$, show further that

$$\nabla^2 E_z + \beta^2 E_z = -\frac{1}{\epsilon}\left[E_z\frac{\partial^2\Delta\epsilon}{\partial z^2} + \frac{\partial E_z}{\partial z}\frac{\partial\Delta\epsilon}{\partial z}\right]$$

4. Obtain an approximate solution for

$$u_{xx} + u_{yy} - \epsilon^2 u^2 = 0$$

within the unit disc $x^2 + y^2 < 1$ in the xy-plane subject to the condition $u(x, y) = 1$ on $x^2 + y^2 = 1$.

5. Suppose the parameters of the medium filling an electromagnetic cavity are (ϵ, μ, σ) and let the resonant frequency be ω_0. If the medium parameters suffer a minute perturbation to new values $(\epsilon + \Delta\epsilon, \mu + \Delta\mu, \sigma + \Delta\sigma)$, show that the fractional change in the resonant frequency is given by

$$\frac{\omega - \omega_0}{\omega} = -\frac{\iiint [(\Delta\epsilon - j\Delta\sigma/\omega)\mathbf{E}\cdot\mathbf{E}_0 - \Delta\mu\mathbf{H}\cdot\mathbf{H}_0]\,d\tau}{\iiint [(\epsilon - j\sigma/\omega)\mathbf{E}\cdot\mathbf{E}_0 - \mu\mathbf{H}\cdot\mathbf{H}_0]\,d\tau}$$

6. Show that the fractional change in the resonant frequency of a rectangular cavity of volume τ due to introduction of a small dielectric sphere of volume $\Delta\tau$ at its center is given by

$$\frac{\omega - \omega_0}{\omega} = \left(\frac{1 - \epsilon_r}{2 + \epsilon_r}\right)\left(\frac{6\Delta\tau}{\tau}\right)$$

where ϵ_r is the relative permittivity of the dielectric sphere introduced.

7. The dominant resonant frequency of an air-filled cylindrical cavity of radius a is given by

$$\omega_{\text{cyli}} = \frac{2.405c}{2\pi a}$$

and that of a cavity with a base in the form of an equilateral triangle of side $a\sqrt{3}$ is given by

$$\omega_{\text{tri}} = \frac{4\pi c}{3\sqrt{3}a}$$

where c is the velocity of light. These two cross-sections are shown in Figure 6.12. Discuss a perturbational technique that changes the resonant frequency from one to the other as the shape of cross-section changes.

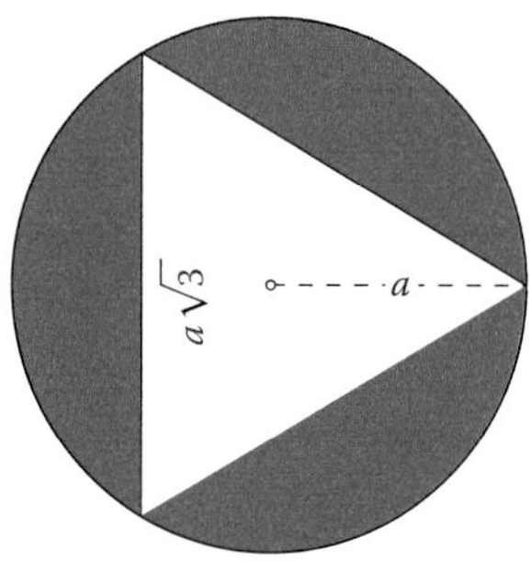

FIGURE 6.12
A circular cavity being reshaped into another with a cross-section like an equilateral triangle.

Bibliography

[1] C. A. Balanis, *Advanced Engineering Electromagnetics*, New York: Wiley, 1989.

[2] K. J. Binns and P. J. Lawrence, *Analysis and Computation of Electric and Magnetic Field Problems*, Oxford: Pergamon Press, 1963.

[3] J. Brown and R. Churchill, *Complex Variables and Applications*, 8th ed., New York: McGraw-Hill, 2008.

[4] D. G. Duffy, *Transform Methods for Solving Partial Differential Equations*, Boca Raton: CRC Press, 1994.

[5] H. J. Eom, *Electromagnetic Wave Theory for Boundary Value Problems*, New York: Springer, 2004.

[6] R. Garg, *Analytical and Computational Methods in Electromagnetics*, Boston: Artech House, 2008.

[7] R. N. Ghose, *Microwave Circuit Theory and Analysis*, New York: McGraw-Hill, 1963.

[8] R. F. Harrington, *Time-Harmonic Electromagnetic Fields*, 8th ed., New York: Wiley-IEEE Press, 2001.

[9] N. N. Lebedev, I. P. Skalskaya, and Y. S. Uflyand, *Worked Problems in Applied Mathematics*, New York: Dover Publications, 1965.

[10] Z. Popovic and B. D. Popovic, *Introductory Electromagnetics*, Upper Saddle River, NJ: Prentice Hall, 2000.

[11] M. N. O. Sadiku, *Numerical Techniques in Electromagnetics*, 3rd ed., Boca Raton: CRC Press, 2009.

[12] M. N. O. Sadiku, *Elements of Electromagnetics*, 6th ed., New York: Oxford University Press, 2014.

[13] F. T. Ulaby, *Applied Electromagnetics*, Upper Saddle River, NJ: Prentice Hall, 1997.

[14] E. Zauderer, *Partial Differential Equations of Applied Mathematics*, New York: John Wiley & Sons, 1983.

Index